IMAGES
of America

INDIAN POINT NUCLEAR POWER PLANT

This is what made Indian Point tick—nuclear fuel assemblies. Each one contained thousands of uranium dioxide pellets inside 204 fuel rods. When allowed to fission, they generated an enormous amount of heat inside the reactor. This one, just removed from the reactor, displays the ghostly Cherenkov radiation, the blue glow caused by charged particles moving faster than the speed of light in water. (Photograph by the author.)

On the Cover: Operator Jim Oakman works the controls at the manipulator console in 1965. This was Indian Point 1's first refueling. The viewing plate to the reactor below is at his feet. Oakman's daughter Madalyn began working at Indian Point as a 17-year-old and had a 38-year career at the plant. She met her husband, Dan Noto, at Indian Point. He had a 35-year career at the plant. (Courtesy of Con Edison.)

On the Back Cover: Pictured is the Indian Point Heritage Project logo. The project was founded and directed by Dr. Michael Conrad of Clarkson University and sponsored by Entergy. Extensive audio interviews with Indian Point employees were recorded. These interviews, as well as photographs and videos, can be found at ipheritage.org. (Courtesy of Joseph DePinho.)

IMAGES
of America

Indian Point Nuclear Power Plant

Brian R. Vangor

ISBN 978-1-4671-0923-9

Published by Arcadia Publishing
Charleston, South Carolina

Printed in the United States of America

Library of Congress Control Number: 2022946527

For all general information, please contact Arcadia Publishing:
Telephone 843-853-2070
Fax 843-853-0044
E-mail sales@arcadiapublishing.com
For customer service and orders:
Toll-Free 1-888-313-2665

Visit us on the Internet at www.arcadiapublishing.com

To my wife, Cathy, my rock—I could not have endured over four decades at the plant or written this book without you.

Contents

ACKNOWLEDGMENTS

A special thank-you goes to Jerry Nappi, Entergy's director of communications. Jerry directed Indian Point's public and media relations through its most difficult years and guided me through my efforts of documenting the legacy of Indian Point. Thank you, Joy Russell, senior vice president and chief commercial officer at Holtec International, longtime friend, and West Point alum. Joy is a leader in nuclear plant dry cask storage and decommissioning technology and a pleasure to all. I thank Allan Drury and Kelly Saunders of Con Edison's Media Relations Department and Ethan Reigelhaupt, NYPA's vice president of corporate communications for their permission in the use of vintage Indian Point photographs.

Joe Goebel, Gail Ruh, Mike Ruh, Rich Jones, Bill Durr, Steve Munoz, Gary Hinrichs, John McAvoy, Bill Lettmoden, and the late Charlie Limoges provided an enormous amount of assistance with their ability to recall events and find former employees. Jon Summers's outage photographs were a vital resource. All had lifelong careers at the plant, and each was a major contributor to the success of Indian Point. From the Village of Buchanan, thank you to George Boyle, history consultant, and Sean Murray, former mayor and current village trustee and Indian Point employee. Both provided vital information and the unfettered use of the Buchanan Local History Room for my research.

Barbara and Wes Gottlock's previous research on Indian Point Park was an invaluable source of information about the park. John Curran, vice president of the Peekskill Museum, provided vintage Indian Point photographs, as did Dr. Carla Lesh, collections manager and digital archivist at the Hudson River Maritime Museum. Lisa Adams Cole shared her amazing collection of Indian Point Park memorabilia with me—thank you to all.

Thank you to my two Saunders Trade and Technical High School teachers Michael and Elaine Richardson (machine design and mathematics), who worked my class hard and prepared us for the rigors of mechanical engineering school yet to come. Finally, thank you to my dad, Robert J. Vangor, tool and die maker, who not only sent me to Saunders (his high school), but also gave me my interest in all things mechanical.

Introduction

Sometime in the mid-1960s, my family and I got into our 1961 Ford Fairlane and drove from our home in Yonkers to Buchanan, New York. My dad wanted to see what this "Indian Point" was all about. I remember a small visitor's center, a big guy with a blue hard hat handing out pamphlets, and then a short movie. The finale was walking to an outdoor deck and seeing the white dome of Indian Point 1. A few years later, I became a serious fan of *Star Trek* and decided to someday become Captain Kirk. I didn't realize how these two events would eventually become intertwined.

Indian Point was a melting pot of extraordinary people with unique characters; some were local, some came up from Con Edison in New York City, and others from the US Navy or local engineering schools. Many employees found the unique nature, complexity, and excitement of the job addictive. It was not a revolving door. People stayed for decades or their entire careers. Homes were bought. Children were born. Sacrifices were made. Lifelong friendships were formed. The stories that could be told are too numerous to recount. It became our "Shangri-la on the Hudson."

Indian Point was 24 hours a day, seven days a week. Those responsible for standing the watch worked Christmas, New Year's Eve, Independence Day, and their kids' birthdays. Operations, Radiation Protection, Chemistry, and Security were always there. Whether I was in a gas station in Peekskill or at the Grand Canyon, I knew that people who were qualified, highly intelligent, and trustworthy were always in the control rooms and elsewhere in the plant. Eventually, my days off would end, and I would return to "the watch." The line would continue, and the story never ended.

I spent 35 years in the Operations Department, all that time working in and around the Indian Point 3 (IP3) control room. Life in the control room was unlike any other. Three of us were "in the box" for 12 hours. We could not leave unless properly relieved. There were no windows or TVs. Logs were taken and tests were performed. Day shifts were a constant stream of visitors and unending phone calls. It took some time for new operators to become comfortable around the hundreds of gauges, switches, knobs, and lights of the control room's seafoam green panels. Eventually, they did. We were isolated, and perhaps unlike life elsewhere, everything made sense in that room. We knew what to expect. After decades in the control room, some came to depend on those green panels. There was a very continuous and routine aspect to it all. If anything out of the ordinary happened, we were well trained to handle it.

When not in the control room, many long nights were spent walking the plant and gazing at the equipment. For an engineer, it was like a candy store. The amount of thought that went into Indian Point always astounded me—even after 40 years. Not only was the plant built to protect the public, it was also built to protect us. For decades, modifications were made, safety evaluations were written, pumps and valves were repacked, transmitters were calibrated, and outages were planned. But through it all, I always wondered how much the plant itself felt our impact. The designers of Indian Point were more clever than we knew. I sometimes thought that the plant ran despite us constantly swirling around it. But we were the stewards of this incredible machine, obligated by law, our technical judgment, our good conscience, and our love of the plant to do the things that we did.

I realize that this book is written from an IP3 Operations perspective. It is where I am from and where my camera was. In addition, quality photographs do not exist of all groups. But Indian Point was not only Operations, Radiation Protection, Chemistry, and Security. Other departments made meaningful contributions and had their own histories and stories. Maintenance and Instrumentation & Control kept the plants operating through a constant process of corrective and preventative maintenance. Engineering monitored each system and searched for ways to improve them. The Performance group tested numerous systems and measured plant efficiency. Work Control built schedules and prepared work packages. Licensing evaluated changes to our technical specifications and answered questions from the Nuclear Regulatory Commission. Rad Waste processed liquids and sluiced and shipped spent resin. The warehouse and Procurement inspected and stored thousands of parts. Quality Services ensured that all work was done correctly. Training provided initial and requalification programs, examined students, and evaluated the results. Emergency Planning ran day-long drills for the entire staff. Safety identified hazards and evaluated industry events. Workers from the local union halls were a constant fixture at Indian Point. As the turbines turned, everyone did their jobs and did them well.

Someone recently asked me what my one takeaway would be from working at Indian Point. My response was that somehow, Indian Point ended up with some of the best and brightest that the area had to offer. They all seemed to migrate to the plant knowing that their abilities could be brought to fruition. They worked long hours, stayed all night to resolve emergent issues, had few days off during long outages, and sometimes worked in very physically demanding environments. Through it all, they applied world-class technical knowledge, unyielding conservatism, and a steadfast desire to do the right thing. I was extremely fortunate and honored to have been part of this Indian Point group and the first wave of commercial nuclear power in this country. I will forever treasure my friends, my countless memories, and the knowledge of the challenges we overcame to reach our goals. I wish the same good fortune to those of the second wave—small modular reactors.

My goal in writing this book was to pay homage to Indian Point and its people. I wanted to make this book technical enough so that Indian Point veterans would find it interesting, but general enough so that "non-nukes" would achieve a much better understanding of how it all worked. I hope I succeeded.

In the end, Indian Point became as much of a Hudson River story as the steamships and park that preceded it. Thousands of people strived for their entire careers to provide a service and make a living in this highly technical, regulated, and labor-intensive endeavor. They succeeded beyond expectations. They set a world record. Through it all, they became the best in the world at what they did.

One

Indian Point Park

In the days when steamships ruled the Hudson River, the Hudson River Day Line was the most prominent carrier. Boasting five large steamships in 1921, the Day Line provided passenger service from New York City to Albany and several stops in between. However, revenues were lost as a significant number of passengers went ashore at the popular Bear Mountain each day. The Day Line finally decided to have a park of its own and enjoy all-day captive customers. In 1923, the Day Line purchased 320 acres on the river just south of Peekskill, the site of the former Bonner Brick Company and, most recently, Michael's Farm. The property was converted into a park that included picnic areas, quiet trails, playgrounds, a cafeteria, a beach, and a swimming area. The park opened on June 26, 1923, and was given the name Indian Point. The company knew that the Native American Kitchawank nation had lived in the area and assumed that the name would attract its younger passengers. In the following years, the park flourished. A 100-foot-by-150-foot pool with bathhouse and lockers was added in 1929. Indian Point Park was a welcome and pleasant escape from the hustle and bustle of New York City.

By the late 1940s, the automobile provided many more options to those seeking peace and quiet. The Day Line began losing revenue. In 1948, it decided to go out of business. The property, including Indian Point Park, was auctioned off in 1949 and then resold to Emanuel Kelmans, who reopened it in May 1950. Kelmans turned the park into a full-fledged amusement park with rides and concessions. By 1952, a beer hall, miniature train, Ferris wheel, and miniature golf were among many other attractions that were added. The Westchester County Fair was held at Indian Point Park in 1952 and 1953. Kelmans sold the property to Con Edison on October 8, 1954, for $250,000. He kept the park in operation for two more years, allowing his concessionaires to find other venues. This idyllic landscape on the shores of the Hudson River would soon change dramatically.

In the very early days of Indian Point Park, the Day Line steamer *DeWitt Clinton* approaches from the south. The park had two large piers and a dock for speedboat rides. The open-air assembly hall (or dance hall) welcomed visitors to the pristine grounds. (Courtesy of Donald C. Ringwald Collection, Hudson River Maritime Museum.)

The bell from the retired steamer *Mary Powell* was situated on the riverfront near the boat landings. The *Mary Powell* was known as "the Queen of the Hudson" because of her speed, grace, and punctuality. The bell was rung each time a ship arrived at the park to alert workers of incoming visitors. Today, the bell is on display at the New York Historical Society Museum in Manhattan. (Courtesy of Village of Buchanan William J. Burke Historical Room.)

Another frequent visitor to Indian Point Park was the Day Line steamer *Peter Stuyvesant*. This 270-foot steel-hulled vessel was built for the Day Line in 1927 at a cost of $728,000. She operated for the company up until 1962. (Courtesy of Village of Buchanan William J. Burke Historical Room.)

This is a view of the popular pool at Indian Point Park. It measured 100 feet by 150 feet and was added in 1929. The drained pool survived the construction of Indian Point 1 and was eventually filled in to make a gravel parking lot for Unit 3, known as the "contractor's lot." (Courtesy of Village of Buchanan William J. Burke Historical Room.)

Ten-year-old Marion Ackerman has arrived at Indian Point Park with her parents, Henry and Philomena, aboard the *Hendrick Hudson* in 1938. This Day Line steamship was built in 1906 at the Thomas S. Marvel shipyard in Newburgh, New York. Marion's son Gary Hinrichs would have a 37-year career in engineering at the Indian Point Nuclear Power Plant. (Courtesy of Gary Hinrichs.)

Marion Ackerman poses with her mother in front of the Native American teepees that greeted visitors to the park. Chief Eagle Plume and two women demonstrated Native American traditions and sold souvenirs such as knives, headdresses, and blankets. This presentation was aimed at giving credence to the name and nature of the park. (Courtesy of Gary Hinrichs.)

The Day Line auctioned off the park property in 1949. It was resold to Emanuel J. Kelmans, who reopened the park in May 1950. Seen here at the ribbon-cutting ceremony is Buchanan mayor William J. Burke (second from left) and Emanuel Kelmans (light suit at center). (Courtesy of Village of Buchanan William J. Burke Historical Room.)

Under his ownership, Emanuel Kelmans changed the character of Indian Point Park from a pleasure park to an amusement park. By 1952, the park had many added amusements and 50 concessions. Kelmans also had an arrangement with Coca-Cola so that only its products were sold at the park. (Courtesy of Village of Buchanan William J. Burke Historical Room.)

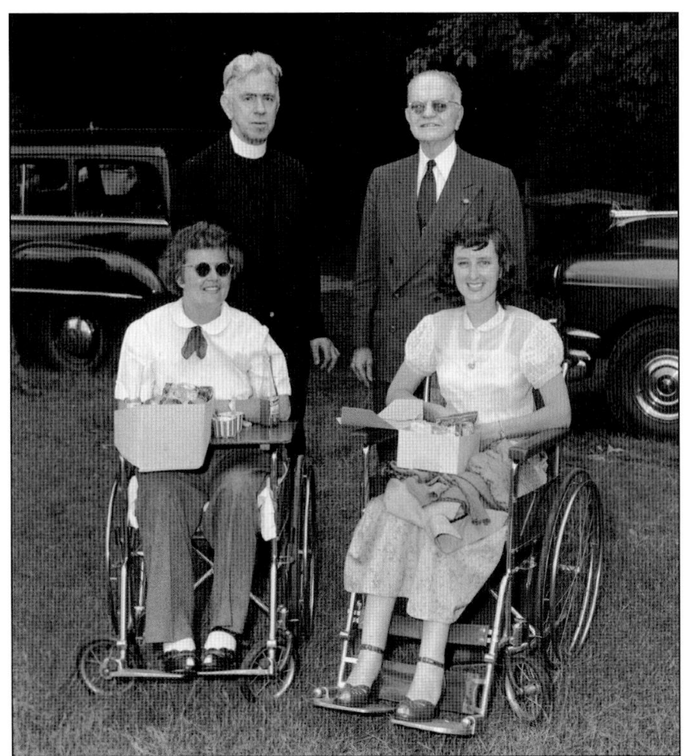

Retired admiral William F. "Bull" Halsey (top right) accompanied a group of 1,405 physically handicapped individuals to Indian Point Park on June 11, 1952. This was the 23rd annual outing to the park sponsored by the Institute for the Crippled and Disabled. Admiral Halsey presented graduation awards for rehabilitation achievements on the steamship *Peter Stuyvesant* during the excursion from New York City. (Courtesy of Village of Buchanan William J. Burke Historical Room.)

This image was taken on July 4, 1952. The upcoming Westchester County Fair was being held at Indian Point Park for the first time. This motorized tram ride transported patrons from the parking lot and throughout the park. Dodgem was a bumper car ride for 25¢. (Courtesy of Village of Buchanan William J. Burke Historical Room.)

Speed boat rides were a popular attraction at the park. In this July 1952 view, a portion of the 189 ships of the Hudson River Reserve Fleet is seen at anchor. The ships were moored in rows of 10 that stretched nearly three miles. They were slowly removed and auctioned off in batches, with the last ship removed on July 8, 1971. (Courtesy of Village of Buchanan William J. Burke Historical Room.)

GOOD AFTERNOON

THE EVENING STAR

THE WEATHER

VOL. XXXII—NO. 256 PHONE PE 7-1200 PEEKSKILL, N. Y., FRIDAY, OCTOBER 8, 1954 14 PAGES FIVE CENTS

ATOMIC POWER PLANT MAY BE BUILT AT INDIAN POINT

Vassar Senior, Mill Hand Die; 9 Others Hurt

Uranium Prospecting At Camp Smith To Commence Next Week

19-Year-Old Music College Student 7th Polio Case Here

CON EDISON PURCHASES INDIAN POINT

Con Edison Buys Park For Future Expansion

Suitability of Site for World's First Peace A-Plant Being Studied

The 242-acre Indian Point Amusement Park on the Hudson River in Buchanan has been acquired by the Consolidated Edison Co. On the site may be built the world's first atomic plant to generate electric power.

Louis A. Scofield, Con Edison vice-president, confirmed today that the purchase had been made from the Indian Point Park Corp. as the site of an electric generating station.

"The suitability of the site for an atomic power plant is under study," said Mr. Scofield.

Indian Point Park To Run 15 Years; Sold For $250,000

The front page of Peekskill's *Evening Star* on October 8, 1954, announced the possibility of an atomic power plant being built at Indian Point. The 242-acre park property had just been sold to Con Edison. The park would remain in operation for another two years, allowing concessionaires to find other venues. (Courtesy of the Peekskill Museum.)

Public Information Bureau
CONSOLIDATED EDISON COMPANY OF NEW YORK, INC.
4 Irving Place, New York 3, N. Y.
Tel: GRamercy 3-5600, Ext. 4113, 4116

March 28, 1955

FOR RELEASE TUESDAY, MARCH 29

Consolidated Edison today (Tuesday, March 29) released details of the $55,000,000 nuclear steam electric generating station it seeks to build at Indian Point in Buchanan, N. Y.

The company applied last week (Tuesday, March 22) to the Atomic Energy Commission for authority to build and operate the reactor portion of the plant.

In its application, the company says it wants to build a pressurized water thorium-uranium converter reactor. The pressurized water concept, the company states has been demonstrated both as to safety and dependability in the naval reactor program.

In the converter design, non-fissionable thorium is converted into a fissionable form of uranium (U-233) inside the reactor. Since the expense of handling fuels is an important element in operating costs, the company expects the use of thorium to result in savings.

For reasons of economy, the steam leaving the nuclear boilers will be heated additionally in an oil-fired superheater. This will result in higher steam temperature and increase the capacity of the plant while reducing production costs at the same time.

Electric capacity of the station is calculated at 236,000 kilowatts, a figure which is subject to revision because of the novel nature of the design. The power produced from the atom's energy will be fed into the network of power lines supplying Westchester County and New York City.

# # # # #

This March 28, 1955, public information release from Con Edison provided more details on the proposed "nuclear steam electric generating station" that it was planning to build. Its application to the Atomic Energy Commission highlights the already proven pressurized water reactor concept and the economic advantages of the use of the thorium-uranium converter design. (Courtesy of Con Edison.)

Pictured is a poster from the Indian Point Park collection of Lisa Adams Cole. The poster advertises a two-day Gypsy Tour event at the park sponsored by the Metropolitan Harley-Davidson Dealers Association. Other items in her collection include signs, pennants, pottery, cutlery, beaded necklaces, charm bracelets, patches, coins, medallions, and advertisements. (Courtesy of Lisa Adams Cole.)

In 1956, the final year of the park, the Christiani Brothers Circus was held at Indian Point Park. Its original destination in nearby Cortlandt was deemed too small for the large circus tent. Thousands attended the circus, including 150 orphans from the St. Joseph's Home in Peekskill. (Courtesy of Village of Buchanan William J. Burke Historical Room.)

This lazy, hazy summertime view of Indian Point Park in the mid-1950s shows the arcade and concession area. The Ferris wheel can be seen in the background. This was life in a much simpler time, with no backpacks, water bottles, strollers, digital cameras, or cell phones. (Courtesy of Village of Buchanan William J. Burke Historical Room.)

This is a 2022 view of the concrete staircase that ascended to the back of the pool house from the lower portion of the park. These stairs, near a little-used roadway on the Indian Point Nuclear Power Plant property, have been the only link to the bygone park for decades. (Photograph by the author.)

Two

Nuclear Power Comes to the Hudson Valley

By the 1950s, the demand for electricity grew rapidly as New York City and Westchester County populations swelled. The Consolidated Edison Company of New York anticipated a 300 percent increase in the demand for electricity in Westchester County alone over the next 16 years. In February 1955, Con Edison announced that it planned to build an atomic power plant at Indian Point. It was to be the first totally privately funded project in the United States following the Atomic Energy Act of 1954, which opened the door for the private, peaceful use of the atom.

Physical construction of Indian Point 1 (IP1) began in December 1956. It had an output of 275 megawatts electric—approximately 60 percent of that from the reactor plant, with the remaining 40 percent from an oil-fired superheater. Babcock & Wilcox was the primary contractor and was responsible for most of the major components. One of those components was the nuclear reactor vessel. It was 255 tons, 40 feet long, and contained 120 nuclear fuel assemblies. The reactor would be part of a four loop pressurized water system. Con Edison was the architect-engineer, with Vitro Engineering acting as consultant. IP1 was completed in 1962 at a cost of $125 million. It went critical for the first time on August 2, 1962, at 5:42 p.m. After extensive testing, it was tied to the electrical grid on September 16 and reached full power on January 25, 1963.

On October 31, 1974, at 11:00 p.m., IP1 was shut down since it was not in compliance with the Atomic Energy Commission's new criteria for an emergency core cooling system. In January 1976, all fuel was offloaded from the IP1 reactor when Con Edison decided that the cost of modifying the plant to meet the new requirements for an emergency core cooling system could not be justified. IP1 remains today situated between Indian Point 2 (IP2) and Indian Point 3 (IP3). Various IP1 auxiliary systems provided support to IP2 for decades.

GOOD AFTERNOON

THE EVENING STAR

VOL. XXXII—NO. 188 PHONE PE 7-1200 PEEKSKILL, N. Y., TUESDAY, DECEMBER 7, 1954 12 PAGES FIVE CENTS

INDIAN PT. POWER PLANT TO COST $300,000,000

Yorktowner Found Dead In Garage

6-Units to Be Built Between 1957-1975

Executives of the Consolidated Edison Company of New York last night unfolded detailed plans for the construction of a six-unit electric power generating station in Buchanan to serve the needs of Westchester at an ultimate cost of $300,000,000 between 1957 and 1975.

Elaborating on its original announcement last October, when the Indian Point parkland and adjacent Blackley, King and Jacoby tracts were acquired, top-ranking officials of the public utility firm announced at a public hearing on the proposed zoning change for the site that the first unit would cost a minimum of $50,000,000.

The power plant, which may be the first in the nation to use atomic-energy, would serve all of Westchester except a northeast portion where power is furnished by another company and has interconnections with Central Hudson Gas and Electric Company and the Niagara Mohawk Corporation.

The front page of the *Evening Star* on December 7, 1954, touts a $300 million price tag for Indian Point and a plan to build six units at the site. The first unit, IP1, was projected to cost $50 million. (Courtesy of the Peekskill Museum.)

Buchanan mayor William J. Burke poses in a Poirier & McLane bulldozer prior to the start of construction of IP1. Mayor Burke served in his office for 22 years and was instrumental in the property sale to Con Edison. Poirier & McLane was a heavy construction firm headquartered in Yonkers and one of the major contractors used to build the plant. (Courtesy of Con Edison.)

The construction of IP1 begins. Still visible is the ball field, pool, and dance hall from Indian Point Park (top). The containment structure of IP1 would eventually house the reactor and nuclear boilers. It was a 160-foot diameter metal sphere surrounded by a concrete radiation shield. Later, IP2 and IP3 would be concrete domes with an interior metal liner. (Courtesy of Con Edison.)

Senate committee members tour the IP1 construction site on May 23, 1959. Behind the group, the containment sphere can be seen protruding from above the unfinished concrete radiation shield. The erection of the turbine building is in progress on the left. (Courtesy of Con Edison.)

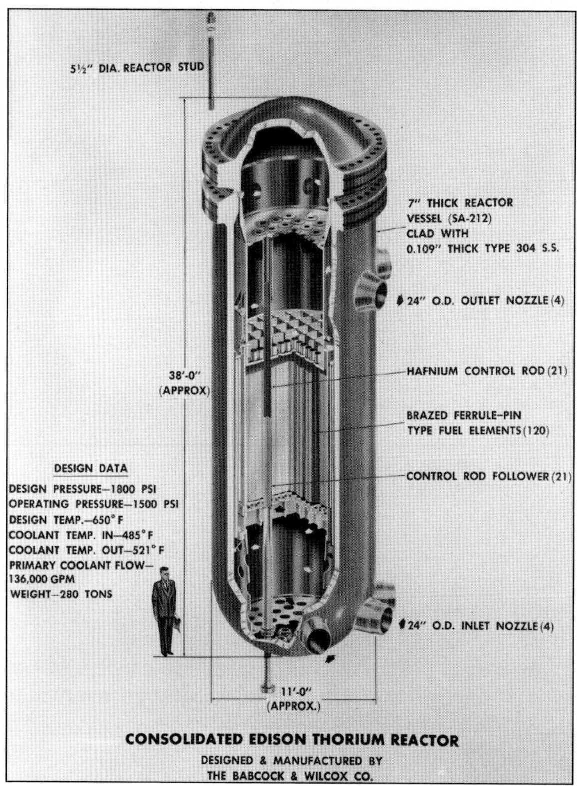

On May 3, 1960, the IP1 reactor vessel was moved into the containment sphere. Built by Babcock & Wilcox in Barberton, Ohio, the 255-ton "nuclear furnace" measured 39 feet, 11 inches long with an inside diameter of 9 feet 9 inches. The shell thickness was 6.9375 inches. The outlet nozzles are seen at the top, and the inlet nozzles are at the bottom. Fifty 5.5-inch diameter bolts held the reactor head onto the vessel. (Courtesy of Con Edison.)

This is a detailed view of the IP1 thorium reactor. The initial core (Core A) was comprised of fuel assemblies of thorium-uranium. IP1 was the first commercial nuclear plant to utilize thorium in a pressurized water reactor. The idea was to convert thorium 232 into uranium 233, hence "converter reactor." Thorium was thought to have distinct advantages in fuel efficiency, but its extraction from the spent fuel proved too costly. (Courtesy of Con Edison.)

To overcome the outward thrust, almost 300 miles of steel wire were wrapped around the radiation shield before any of the precast concrete dome ribs were installed on top. In this April 1959 photograph, the wrapping cart is shown during one of its approximately 2,700 trips around the structure. (Courtesy of Con Edison.)

The 108-foot elevation was the "top floor" inside the containment sphere. The reactor and the storage pool were all located below this elevation and partially accessible through removable floor plugs and hatches. In this 2020 view, much of the original refueling equipment is long gone. The polar crane was used to lift the reactor head, reactor internals, and other heavy equipment. It had a rotatable 50-ton main hook and a five-ton auxiliary hook. (Photograph by the author).

This view is looking down from the 108-foot elevation into the reactor internals storage pool inside the containment sphere. The reactor "upper grid" is seen stored in the lower cavity. The date was January 30, 1962, prior to initial operation. Four new fuel assemblies can be seen in the transfer basket rotating mechanism against the lower cavity wall. The reactor itself is out of view above. (Courtesy of Con Edison.)

In early 1962, technicians lower the upper grid into the reactor vessel. The head grid handling tool is attached to the polar crane above. In conjunction with extension shafts, it used air-operated pistons to remotely connect to reactor internal components. It was manually connected to the head lifting lugs with clevis pins. (Courtesy of Con Edison).

Technicians inspect the IP1 reactor vessel head in its stand on the 108-foot elevation of the containment sphere. The polar crane and head grid handling tool, used to lift the head, are above and to the right. The extension shafts and elevator structure are on the left. (Courtesy of Con Edison.)

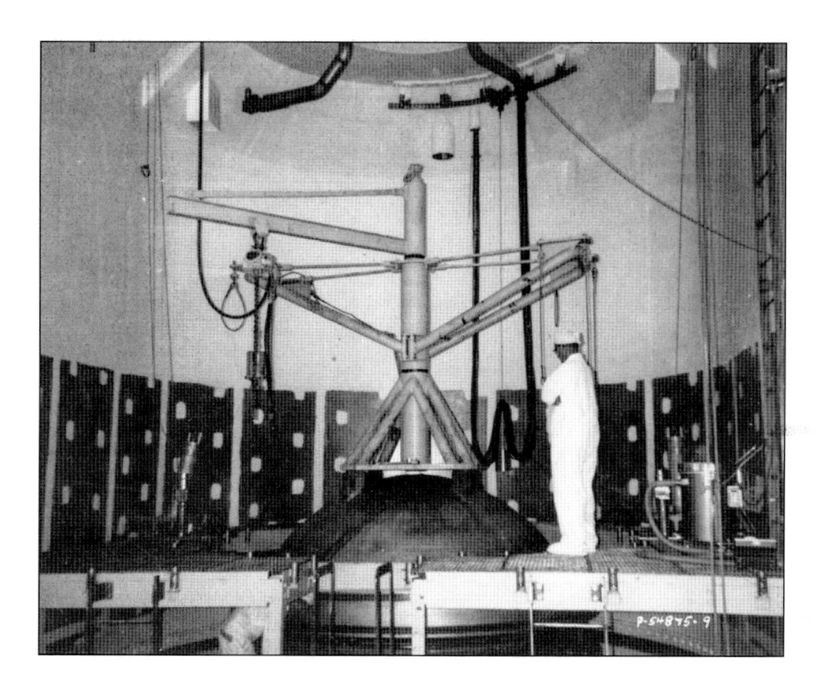

The reactor vessel head is seen placed on the reactor vessel. The structure attached to the top of the head is the stud tensioner support rig. It was used to hold the two stud tensioning machines that either loosened or tightened the 50 head studs that secured the head in place. (Courtesy of Con Edison.)

Shown here is one of the four "nuclear boilers" or steam generators. Water heated by the reactor entered from the left and traveled through 811 U-shaped tubes within the boiler and then exited through the two pipes at upper right. Secondary water flowing around the tubes was turned into steam and routed to the superheater and ultimately the main turbine. (Courtesy of Con Edison.)

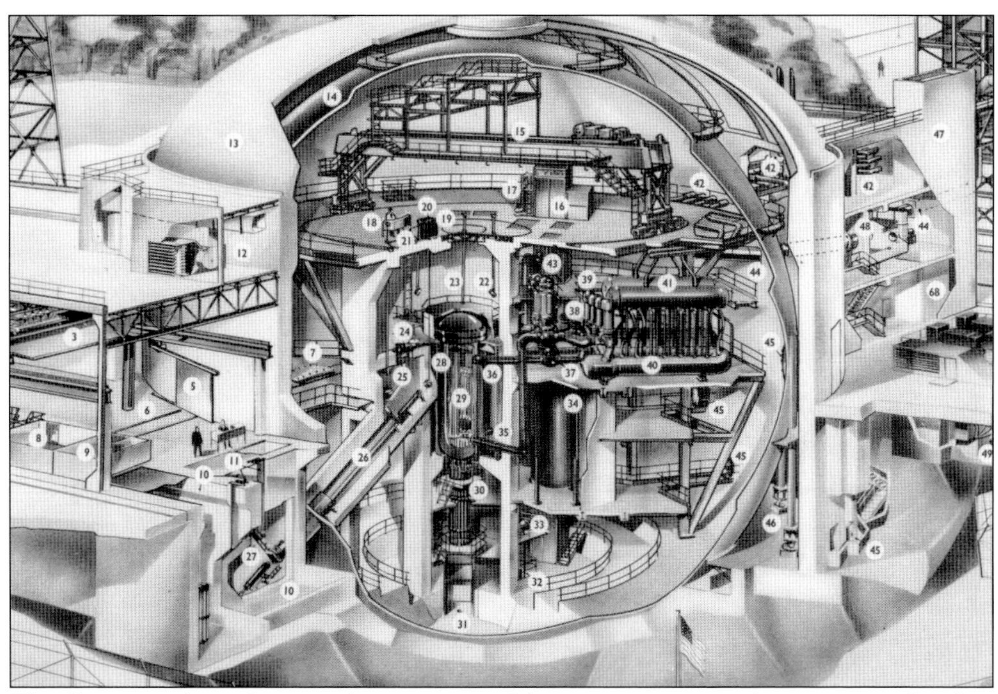

This illustration shows the general arrangement of the reactor, reactor coolant pumps, and nuclear boilers inside the containment sphere. The polar crane is seen on the top (108-foot) elevation. The control rod drive mechanisms are below the reactor vessel. The fuel transfer tube is seen extending at a 30-degree angle down into the fuel handling building. (Courtesy of Nuclear Engineering International.)

IP1 was completed in 1962 at a cost of $125 million. It went critical for the first time on August 2 at 5:42 p.m. and was tied to the electrical grid on September 16. It had an output of approximately 275 MWe (megawatt electrical)—about 60 percent of that from the nuclear reactor, with the remaining 40 percent from the oil-fired superheater. The pool and baseball field can still be seen. (Courtesy of Con Edison.)

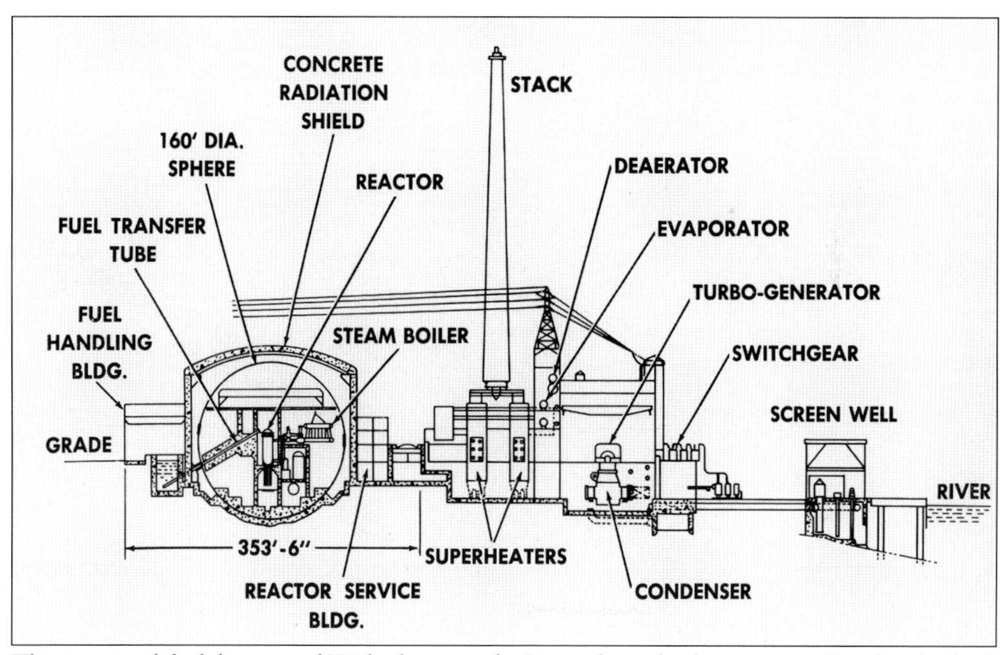

This is a simplified diagram of IP1 looking south. Steam from the four steam or "nuclear boilers" at about 450 degrees was superheated in two oil-fired superheaters to 1,000 degrees. These units, supplied by Foster-Wheeler, were fed No. 6 fuel oil from two 2.5 million-gallon storage tanks up the hillside from the plant. (Courtesy of Con Edison.)

These two photographs show how the plant changed over time. Above, in the mid-1970s, general watch foreman Jerry Walden looks over from the IP1 control room into the IP2 control room. The IP2 "flight panel" is on the left. During construction of IP2, a wall existed between the two control rooms. It was removed when IP2 was completed. Both units ran concurrently for only three months in 1974. Below is the same view in 2021. Modern technology has resulted in many changes, including computers, digital displays, modern communications equipment, and the addition of numerous alarm "cans" on the IP2 flight panel. (Above, courtesy of Con Edison; below, photograph by the author.)

This curved hallway on the 53-foot elevation that skirted the Unit 1 containment building was commonly known as the "banana hallway." It connected the Nuclear Service Building to the Chemical Systems Building. The containment building is on the right. The porcelain tiles on the left were a yellowish beige color, giving the hallway its name. (Photograph by the author.)

This is a view of the "chem systems control room"—many decades removed from serving its original function. It was separate from the main IP1 control room and located in the radiologically controlled area of the plant. From here, operating mechanics controlled Chemical Systems Building and sphere ventilation, waste gas systems, and water reprocessing. (Photograph by the author.)

This is the space between the containment sphere and concrete radiation shield known as the "annulus." The personnel airlock, equipment hatch, two large removable hatches, and six 4-foot diameter personnel escape hatches were located here. Numerous staircases and platforms allowed access to all levels of the sphere. (Photograph by the author.)

The one personnel airlock was on the 57-foot elevation of the containment sphere. The mechanism only allowed one door to be opened at a time, thus maintaining the airtight integrity of the sphere. This view is from the Nuclear Services Building looking into the containment sphere. (Photograph by the author.)

The 50-ton reactor head is being lifted from on top of the reactor vessel. The head grid handling tool is attached to the head, which is in turn attached to the polar crane via extension shafts. Technicians monitored radiation levels as the lift occurred. (Courtesy of Con Edison.)

Indian Point 1 has been shut down for over 45 years in this February 5, 2020, view of the 50-ton IP1 reactor vessel head. The 50 studs and nuts that held the head in place are clearly seen, as are the four attachment points for lifting the head. (Photograph by the author.)

On the 108-foot elevation, technicians walk the extension shafts connected to the reactor head below along the slot in the floor. When they reached the 13.5-foot diameter opening in the floor (partially visible at bottom), the head was lifted through it and placed in its stand on the 108-foot elevation. (Courtesy of Con Edison.)

The head grid handling tool is attached to the head and the polar crane above, as can clearly be seen. A technician signals upward movement to the polar crane operator. Following head removal, the storage pool was filled with water to an elevation of 105 feet for all movement of reactor internals and nuclear fuel. (Courtesy of Con Edison.)

In this view of the 108-foot elevation, the reactor head is sitting atop its stand. The four lifting lugs on top of the head are where the head grid handling tool connects to the head. The four extension shafts are clearly seen. The head studs and elevator are to the right. (Courtesy of Con Edison.)

Much of the refueling system can be seen in this view of the 108-foot elevation. The "manipulator," which moves the fuel into and out of the reactor, is at the end of the J-shaped slot on which it rides. On the short end of the "J" is where it moves the fuel basket into or out of the transfer tube to the fuel handling building. (Courtesy of Con Edison.)

The IP1 reactor core contained 120 fuel assemblies. Each assembly consisted of 195 one-third-inch diameter stainless steel fuel rods that contained hundreds of thorium and later enriched uranium dioxide fuel pellets. Each assembly was approximately eight and a half feet long and about six inches square. In shipments of four, the assemblies made the 518-mile trip from the Babcock & Wilcox Nuclear Facilities Plant in Lynchburg, Virginia. (Courtesy of Con Edison.)

A technician operates the manipulator crane on the 108-foot elevation of the containment sphere. This machine was a nine-and-a-half-foot turntable, 27 feet over the reactor vessel. It positioned a remote fuel handling tool and grapple over specific core components or transfer baskets. Most of the components of the fuel transfer system were built by the United Shoe Machinery Company of Massachusetts. (Courtesy of Con Edison.)

In the above 1972 view from the floor near the reactor head, the round manipulator turntable area is seen directly above the reactor vessel (at bottom) as well as the slot that extends to the transfer tube upender basket location. The photograph below is a modern-day view of the remaining portion of the "J" slot and reactor head concrete floor plug. This round 23-ton floor plug and transition piece had to be removed to lift the reactor head up to the 108-foot elevation. When at the box at the end of the "J" slot, the remote fuel handling tool was directly over the transfer tube upender basket and in position to raise or lower a fuel basket. (Above, courtesy of Con Edison; below, photograph by the author.)

The left view looks straight down through a 30-inch gate valve in the floor of the storage pool and shows the upper end of the fuel transfer tube. An upender rotated the fuel basket to a vertical position. Fuel baskets were moved through the 30-degree angled, 61-foot transfer tube down to the fuel handling building by means of a shuttle car and cable pulley system. Below, the lower end of the transfer tube had a similar mechanism. Operators on the 70-foot elevation of the fuel handling building moved fuel here using a manual tool, grapple, and overhead crane. Fuel was moved through a series of underwater channels, which allowed access to a west and east storage pool, a failed fuel pool, a disassembly pool, and a cask load pool. (Left, photograph by the author; below, courtesy of Con Edison.)

At IP1, the first core (Core A) was a thorium core. It was offloaded from the reactor in December 1965. Core B, installed in early 1966, was a uranium core. Through four additional refuelings, it operated until the final shutdown in October 1974. From 1966 to 1970, a total of 244 fuel assemblies were shipped via a dry cask method to West Valley, New York, for reprocessing. (Courtesy of Con Edison.)

On February 4, 1966, the first of two fuel assemblies are being loaded underwater into a transfer cask in the cask load pit. IP1 was the first commercial US nuclear plant to have its fuel reprocessed. Seventeen tons of stainless-steel clad thorium and uranium oxide pellets from IP1 were reprocessed at West Valley. (Courtesy of Con Edison.)

Seen here are photographs of the IP1 reactor internals storage pool taken 58 years apart. The photograph on this page was taken in 1962 prior to the initial operation of IP1. Here, with the head removed, an operator on the manipulator (top) is indexing the machine to each core location. An operator standing atop the reactor vessel signals to assist in positioning. (Courtesy of Con Edison.)

This is a 2020 view of the previous scene. The reactor was defueled on January 30, 1976. The reactor vessel head was installed, and the 50 bolts were tensioned on February 17, 1976. The refueling equipment, including the manipulator, was removed from the 108-foot elevation many years prior. (Photograph by the author.)

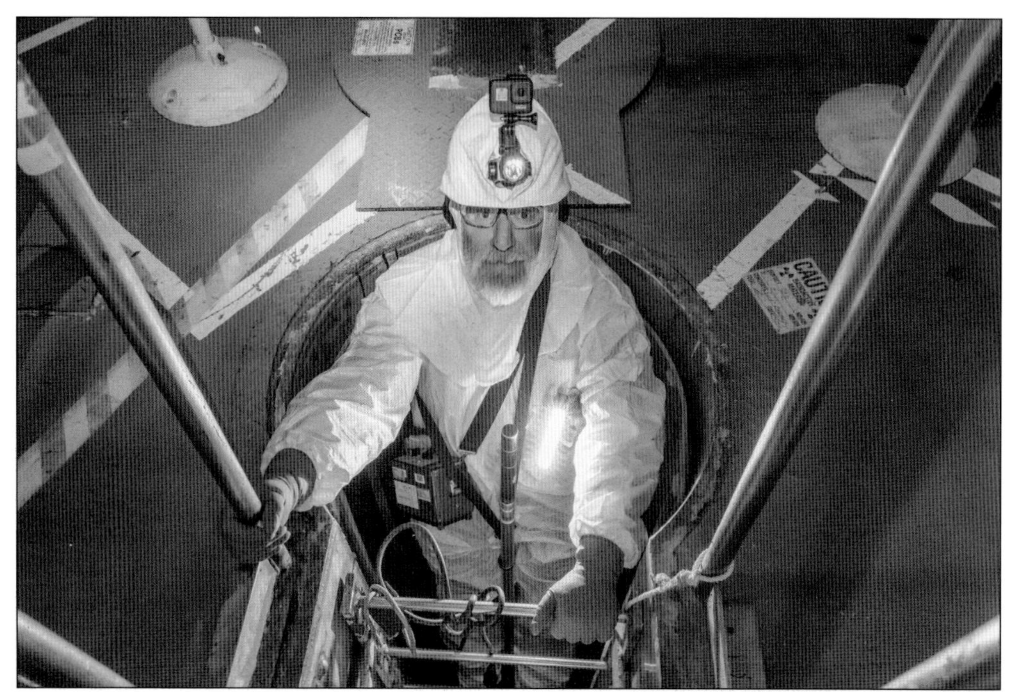

On February 5, 2020, senior radiation protection technician John Dorsey began his descent from the 108-foot elevation down into the (dry) reactor internals storage pool to inspect the reactor head. This was the first time in over 20 years that anyone had entered this area of the plant. (Photograph by the author.)

This is the IP1 control room in 2022. The unit has been shut down for almost 50 years. Many items have been removed and panels have been modified over the decades to include additional controls and instrumentation for IP2. The IP2 control room is to the left. (Photograph by the author.)

Three

Construction of Units 2 and 3

Not long after IP1 achieved commercial operation, Con Edison began planning on two bigger units at the site. Construction of IP2 began in October 1966. Unlike the one-of-a-kind IP1, this plant would be a larger Westinghouse four loop pressurized water reactor. This design was popular at the time with many utilities in the United States. Two years later in November 1968, construction was started on IP3, which would be a sister unit to IP2.

Under subcontract to Westinghouse, United Engineers & Constructors (UE&C) was the initial architect-engineer for IP2 and IP3. Westinghouse would supply the NSSS (nuclear steam supply system)—reactor, steam generators, reactor coolant pumps, and main turbine generator. In 1969, during the middle of the construction of IP2, UE&C gave up certain responsibilities for the completion of both units. A special subsidiary of Westinghouse Electric Corporation, WEDCO, was formed to complete the two plants. The construction of these two units required thousands of individuals to work for years around the clock. Huge amounts of steel, concrete, wood, piping, and wiring, and hundreds of pumps, motors, instruments, and controls were used. Hundreds of engineers produced thousands of drawings to guide the way and bring it all together. Project managers oversaw the daily work. It was a gigantic effort.

After eight years of construction, IP2 achieved commercial operation on August 1, 1974. IP2 was rated at 875 net electrical megawatts. For three months, IP1 and IP2 operated concurrently. On December 31, 1975, feeling the financial burden of building both units, Con Edison sold the unfinished IP3 to the Power Authority of the State of New York (PASNY, later NYPA [New York Port Authority]). Under its new owner, IP3 achieved commercial operation on August 30, 1976. PASNY, however, did not assume operating responsibility until March 1978. IP3 was rated at 965 net electrical megawatts. For the next 24 years, both units would operate under different companies, separated by a fence, and take on slightly different characters. They would be joined again in the 21st century.

Excavation for IP2 began in 1966. The containment building would be in the level area at upper left. The area behind the dump truck would be the transformer yard. The IP1 containment building and three condensate storage tanks are seen in the background. (Courtesy of Con Edison.)

Various structures start to take shape during IP2's construction. From top to bottom are the rectangular-shaped hold-up tank pit (left of IP1 containment), the square-shaped spent fuel pool and circular containment building, the turbine pedestal, the circulating water pump piping, and the intake structure (surrounded by ice). (Courtesy of Con Edison.)

The massive amount of labor and materials that were required to build Indian Point is evident in this 1970 photograph inside the IP3 containment building. The circular wall at bottom left housed the reactor vessel. For seismic considerations, the design specified steel reinforcing bars (rebar) to be set in place prior to pouring concrete in many of Indian Point's walls and structures. (Courtesy of Con Edison.)

Both Indian Point units had two main transformers each. They took the 22,000 volts produced by the main generator and stepped it up to 345,000 volts for transmission on the Con Edison electrical grid. Here, one of them is being delivered by barge to the waterfront dock. Numerous Liberty ships can be seen at anchor on the west side of the river. (Courtesy of Con Edison.)

On August 24, 1970, the IP3 reactor vessel and head arrived by barge at Lents Cove just north of the plant site. Fitting for the Peekskill area, the tugboat *Evening Star* provided the service. The IP2 and IP3 reactor vessels were built by Combustion Engineering in Chattanooga, Tennessee, per Westinghouse Electric Corporation specifications. The vessels themselves weighed approximately 435 tons. The heads weighed 169 tons after being fully assembled. Below, on September 1, 1970, the IP3 reactor vessel on its 85-ton transport skid was slid off the barge, loaded onto a crawler-transporter, and moved up Broadway and into the site at Bleakley Avenue. The vessel heads were moved separately. (Both, courtesy of Con Edison.)

The eight steam generators arrived by barge at Lents Cove (above). Gerosa was a heavy-moving crane company located in the Bronx. Boilermakers from Local 5 performed the work. Below, the lower section of one makes the turn into the site at the corner of Broadway and Bleakley Avenue. In 1968, the IP2 steam generators came in two pieces. In 1970, for IP3, they arrived whole. In both photographs, the inlet and outlet nozzles and two manways as well as the four mounting pedestals can be seen at the bottom of the steam generator. On the corner below at right is the restaurant and bar Johnny Rits, very popular with plant personnel. (Both, courtesy of Con Edison.)

In the unique view above, one of IP3's steam generators is being rotated to a vertical position in December 1970. The rebar and steel plate liner of the containment building wall is quite evident. The polar crane was originally rated to make these heavy lifts of the reactor and steam generators during construction. It was later reconfigured and lowered to the 175-ton rating needed for normal operation. Below, in July 1971, another steam generator (probably 32) is being rotated. On the left, 33 and 34 are already installed. A large pile of sandbags is positioned at the bottom to accommodate the rotation of the lower hemispherical head. (Both, courtesy of Con Edison.)

In this rare 1970 view, IP3's reactor vessel and one of its steam generators are at the bottom of the hill leading up to the containment building. The steam generator is being backed up the hill. The reactor vessel (minus its head) was secured to its transport cradle. The gray metal maintenance service building between the two and right below the tree line was still used onsite in 2022. (Courtesy of Con Edison.)

On June 28, 1971, the No. 31 steam generator was lowered through the 95-foot elevation to its final resting position. A special lifting yoke was attached to the polar crane's block to make these lifts. The rebar extending upward out of the floor would support the concrete biological shield wall that was built around the base of each steam generator. (Courtesy of Con Edison.)

The IP2 reactor vessel is seen here in 1968 being slowly moved toward the entrance of the containment building by the crawler-transporter, nameplate banner still in place. This was a slow and laborious process. The intricate rebar construction of the containment building is seen here as well. (Courtesy of Con Edison.)

On January 12, 1971, the IP3 reactor vessel was rotated inside the containment building to a vertical position using its attached skid. Metal plates are in place over the reactor cavity where the vessel would eventually be located. Steam generators 33 and 34 are seen behind the reactor. (Courtesy of Con Edison.)

On May 26, 1971, the IP3 reactor vessel was lowered into the reactor cavity. The IP2 and IP3 reactor vessels were almost 44 feet high with the head installed and had an inside diameter of 14.5 feet. The minimum vessel wall thickness was 8.625 inches. The four cold leg inlet nozzles had an inside diameter of 27.5 inches, while the four hot leg outlet nozzles were 29 inches. (Courtesy of Con Edison.)

The IP2 reactor vessel is being lowered into its final resting spot in the reactor cavity. When fully lowered, the reactor vessel flange (top of the vessel) was at the same elevation as the floor that the workers were standing on (69-foot elevation). Four of the eight reactor coolant nozzles can clearly be seen. (Courtesy of Con Edison.)

In June 1972, workers stand on the IP2 reactor cavity floor (69-foot elevation) next to the open reactor vessel. Two hot leg nozzles (left) and two cold leg nozzles (right) as well as the keyways for the core barrel (bottom) are seen inside the vessel. The two guide studs extending upward guided the core barrel, upper internals, and reactor vessel head into the proper position. (Courtesy of Con Edison.)

The reactor coolant "loops" were huge stainless-steel pipes that connected the reactor to the steam generators (SG) and reactor coolant pumps (RCP). Each reactor had four loops. Here, work on the 31-inch inside diameter "intermediate leg" of Loop 21 (between 21 SG and 21 RCP) is being completed in 1969. (Courtesy of Con Edison.)

In April 1970, workers at the very bottom of the IP2 reactor vessel clean the incore instrumentation nozzles that extend up from the bottom. Detectors at the end of long cables were periodically extended from these nozzles up into the fuel assemblies to measure nuclear power. One of the six massive keyways that aligns the core barrel inside the vessel can be seen behind the workers. (Courtesy of Con Edison.)

This was the construction overlook for IP2. It was northeast of the IP2 containment building. A Honda N600, a predecessor to the Civic, is parked in front on October 12, 1972. This overlook can be seen in the trees at bottom right in the June 1971 photograph below. Con Edison's Energy Education Center would eventually be located near this line of trees. Although nearing completion, IP2 was still three years away from power operation. The IP3 containment building is under construction and surrounded by cranes. The IP2 turbine building was built attached to the IP1 turbine building forming one large open "turbine hall." The IP3 turbine building was separate. The final few Liberty ships remain at anchor. (Both, courtesy of Con Edison.)

From this bottom view (right), the precise machining of the lower internals, or "core barrel," is quite evident as it is removed from the IP2 reactor vessel in 1969. The six alignment keys can be seen at the bottom. This enormous structure was delivered already inside the reactor vessel but was removed so that final connections and inspections of the reactor vessel could be performed. Below, the IP3 core barrel is being lowered into the reactor vessel on June 22, 1972. The upper half of the core barrel housed the upper internals package. The lower half is the thermal shield, where the "core" consisting of 193 fuel assemblies was eventually loaded. (Both, courtesy of Con Edison.)

This view shows the IP3 side of the construction site. The steel framework of the IP3 turbine building at left is almost fully complete, while the containment building and PAB (Primary Auxiliary Building) have just started. The "fab shop" (lower right), where a variety of metal items were built or modified, remained until 2022. (Courtesy of Con Edison.)

During the construction of each containment building, a polar crane was assembled and installed on the circular crane wall within the building. These cranes were used during construction and then served to perform all heavy lifts for maintenance and refueling outages throughout the life of the plant. IP3 is seen here and IP1 is in the background. (Courtesy of Con Edison.)

The construction of the IP3 containment building shows the process of assembling a tightly woven rebar superstructure, which was then encased in concrete. The walls and dome were between three and four feet thick. American nuclear power plant containment buildings were among the strongest built structures on earth. (Courtesy of Con Edison.)

IP2's 80-foot personnel airlock is seen here on August 20, 1969. The airlock was the entry and egress into the containment building and had an outer door (behind the ladder) and an inner door (opposite end of the tube). Both doors were operated by hand wheels on the left and a similar set at the inner door. The mechanism only allowed one door to be open at a time. (Courtesy of Con Edison.)

The IP2 reactor vessel is being prepared for a test fit of the upper internals package on April 22, 1970. The core is empty as no fuel has been loaded. Three of the four hot leg outlet nozzles can be seen just above the core. The 54 holes around the vessel flange are for the studs that will fasten the head to the vessel. (Courtesy of Con Edison.)

The upper internals package is seen here from the reactor cavity below. This structure sits on top of the 193 fuel assemblies in the reactor core. Pins and flow openings for each fuel assembly can be seen in the upper core plate. The 53 control rods, used as one method of controlling core reactivity, were housed in guide tubes above the core plate. (Courtesy of Con Edison.)

Fifty-three control rod drive shafts extended up out of the upper internals, each connecting to a control rod hub down inside the reactor vessel. Each control rod consisted of 20 rodlets fixed to the spider-shaped hub. When the reactor vessel head was set on top of the upper internals, the drive shafts entered up into the control rod drive mechanisms, which in turn could move the control rods up or down. (Courtesy of Con Edison.)

The main feedwater and main steam lines for IP2's 21 and 22 steam generators are seen here on the 95-foot elevation of containment. Feedwater entered the steam generator (smaller line), was heated by the 600-degree reactor coolant inside 3200 U-tubes, and exited as main steam (larger line). The massive whip restraints held the lines in place in the event of a pipe break. (Courtesy of Con Edison.)

The IP2 main turbine generator is taking shape in the turbine hall in 1969. The main generator is near the center. To the upper right are the three low pressure (LP) turbines. Crossover piping from the six moisture separator reheaters to the LP turbines is being assembled. The IP1 turbine generator is out of view to the bottom left. (Courtesy of Con Edison.)

The casing for the IP3 high pressure (HP) turbine is being lowered into place. The blades of the HP turbine can be seen below the casing. An intricate oil system controlled the four control valves, which changed the amount of steam that entered the turbine and ultimately governed the amount of electrical output. (Courtesy of Con Edison.)

Workers carefully monitor the insertion of the main generator rotor into the stator at IP2 in September 1969. When completed, the rotor would be connected to the three low pressure and one high pressure turbines all inline on one shaft. When in operation and spinning at 1,800 RPM, the generator provided 22,000 volts (stepped up to 345,000 volts) to Con Edison's electrical grid. (Courtesy of Con Edison.)

The inspection of the No. 22 main boiler feed pump is in progress at IP2 on April 21, 1971. Both units had two of these steam-driven pumps, which were the driving force for moving huge volumes of feedwater back into the steam generators. Each pump was rated at 16,000 GPM at a discharge pressure of 1,000 PSI. The driving turbine was rated at 8,350 HP at 5,350 RPM. (Courtesy of Con Edison.)

In 1971, the IP2 spent fuel pool was not yet filled with water. Workers on the fuel handling bridge are using the spent fuel handling tool to lift and unwrap new fuel assemblies in preparation for being loaded into the reactor as the initial core. Prior to the first refueling, the pool will be filled with borated water so that the spent fuel from the core can be safely stored within. (Courtesy of Con Edison.)

The manipulator crane lowered and raised fuel assemblies into and out of the reactor core. The tops of the fuel assemblies were 27 feet below the manipulator. A telescoping mast with a "gripper" at the bottom moved the fuel between the core and transfer system, which connected to the spent fuel pool. All movement of the radioactive fuel was performed under at least 23 feet of water. (Courtesy of Con Edison.)

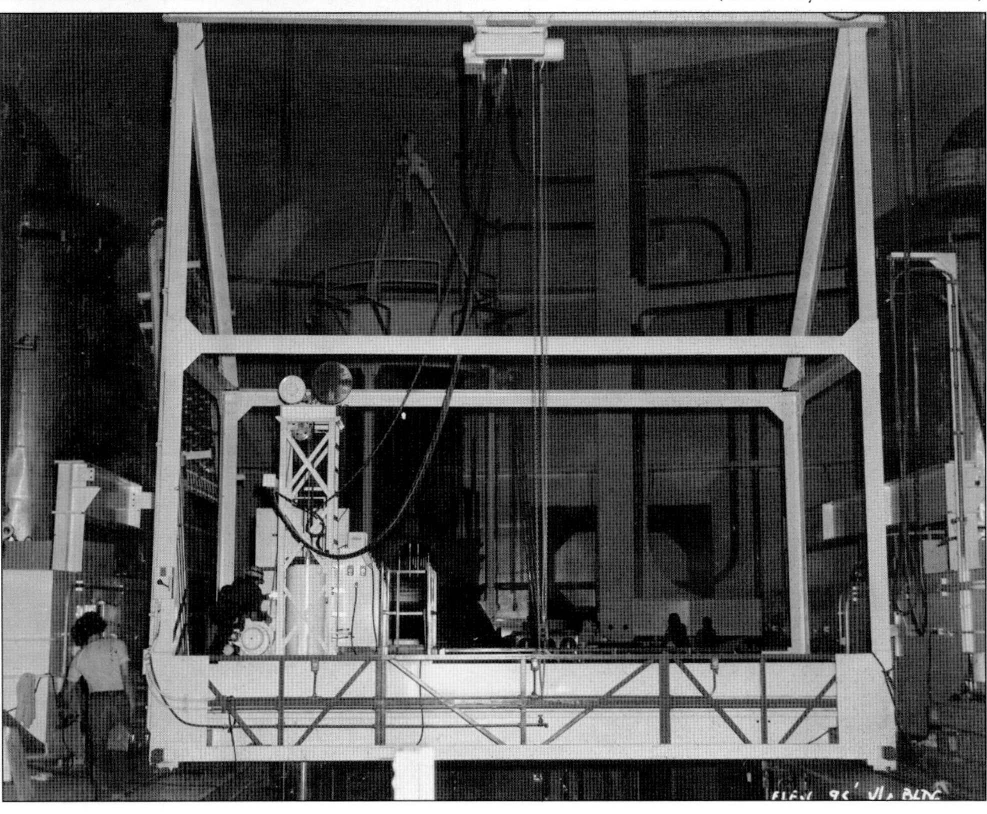

Above, a worker on the 69-foot elevation of the reactor cavity, with a flashlight and clipboard in hand, monitors the lowering of a fuel assembly from the manipulator crane mast into the core. The vessel is partially filled with water. Below, in November 1971, workers using a light on the 69-foot elevation of the reactor cavity inspect the 193 new fuel assemblies of the fully loaded initial core for IP2. The tops of the 13-foot-long assemblies are 11 feet below the workers. The white steel plugs around the vessel kept the stud holes dry after the reactor cavity was filled with water. The bottom of the manipulator crane mast is seen at lower left. (Both, courtesy of Con Edison.)

The 169-ton IP2 reactor vessel head is seen here suspended over the reactor cavity in 1971. The actual closure head is at the very bottom. Above are the control rod drive mechanisms (CRDM), which moved the control rods up or down, and the rod position indication (RPI) coil stacks, which provided control rod position to the control room. (Courtesy of Con Edison.)

In December 1970, the IP2 reactor head is fully seated on the reactor vessel. All 54 closure studs and nuts are installed. They are being tightened by a stud tensioner (machine with the hand wheel on top). Each stud was about five feet long, seven inches in diameter, and weighed approximately 700 pounds. Each nut weighed about 75 pounds. (Courtesy of Con Edison.)

In this 1975 view, IP1 was permanently shut down, IP2 was in operation, and IP3 was still a year away. The IP1 stack was shortened by 80 feet in 1972 in accordance with an analysis of it falling onto the IP2 control room or diesel generators. The IP3 control room can be seen extending out from the bottom left side of the turbine building. (Courtesy of Con Edison.)

During a Family Day at Indian Point in 1970, James and Phyllis Mooney and their children (from left to right) James Jr., Michael, Nancy, Linda, Patricia, and friend JoEllen Stipak pass through the 80-foot airlock into an under-construction IP2 containment building. James Mooney Sr. was the Instrumentation & Control superintendent at all three units. At IP3, James Jr. became a senior reactor operator and engineering manager, and Mike became a work week manager. (Courtesy of Con Edison and the Mooney family.)

Four

Power Operation

It took up to 1,000 people per unit to operate Indian Point safely and efficiently. The major departments included Operations, Engineering, Maintenance, Instrumentation & Control, Radiation Protection, Chemistry, Training, Quality, Safety, Licensing, Administration, and Security. Senior reactor operators, reactor operators, and nuclear plant operators ran the plants from the control rooms and from out in the field. Technicians took chemical samples and performed radiation surveys. The plants were monitored, inspected, cleaned, and adjusted around the clock. Hundreds of preventative and corrective maintenance tasks and periodic tests had to be performed. Engineers evaluated plant performance and designed modifications. Security officers guarded posts, walked established rounds, and monitored hundreds of cameras. Training was a constant. Human performance was observed and evaluated. Audits were performed internally by the Quality Department and externally by the Nuclear Regulatory Commission, the Institute of Nuclear Power Operations, and others.

Indian Point personnel worked long hours. They became used to 12-hour shifts, night shifts, infrequent days off, snowstorms, and more. They missed holidays, Little League games, dance recitals, and family gatherings. Everyone worked together to meet a challenge or solve a problem. Most employees were local and stayed for decades. Over the years, Indian Point became one big family.

In 2000, NYPA sold IP3 to Entergy, a Louisiana-based power producer. The following year, Con Edison sold both IP1 & IP2 to Entergy. Now, after 24 years, the fence came down. Procedures, practices, and personnel were aligned to meet the highest industry standards. Performance improved as both units of the newly named Indian Point Energy Center strived to reach the same goals.

Indian Point developed a rich heritage. Many people in the Hudson Valley either worked at the plant or knew someone who did. Up to three generations of families worked at Indian Point. The plant evolved greatly throughout the decades, changing as the industry changed after Three Mile Island, Chernobyl, and Fukushima. Withstanding constant criticism, it was improved and made more reliable, culminating in a world record run for Indian Point 3 of 753 continuous days of operation, shutting down on April 30, 2021.

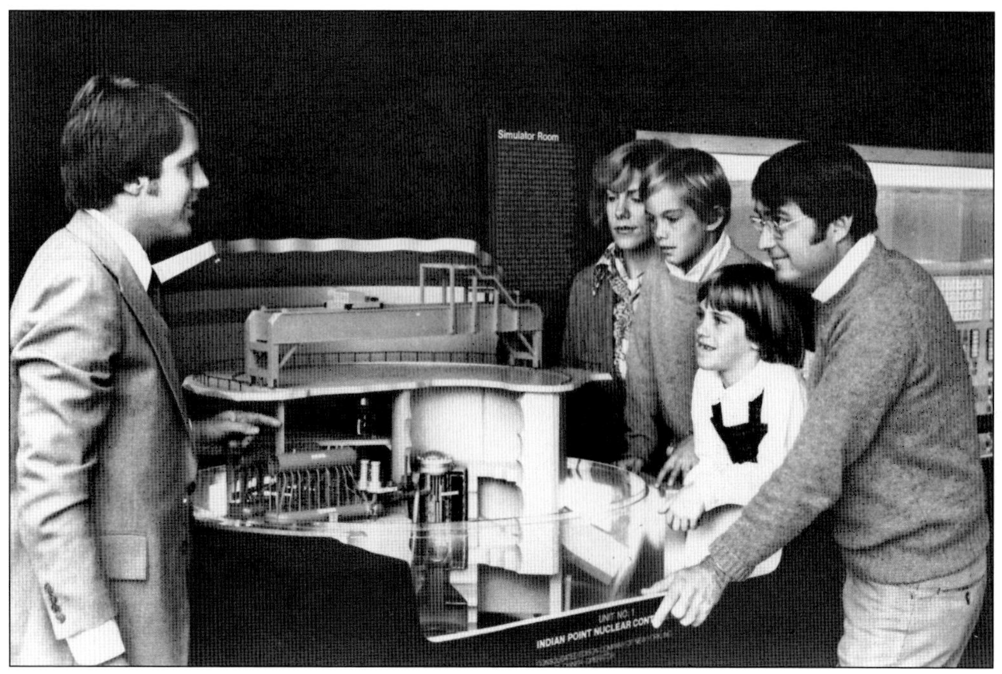

The Energy Education Center (EEC) at Indian Point was built by Con Edison to educate the public about nuclear power. Here, a family listens to a description of the IP1 containment building. The reactor, horizontal nuclear boilers, and polar crane are seen in the model. Behind them is the IP2 control room simulator. (Courtesy of Con Edison.)

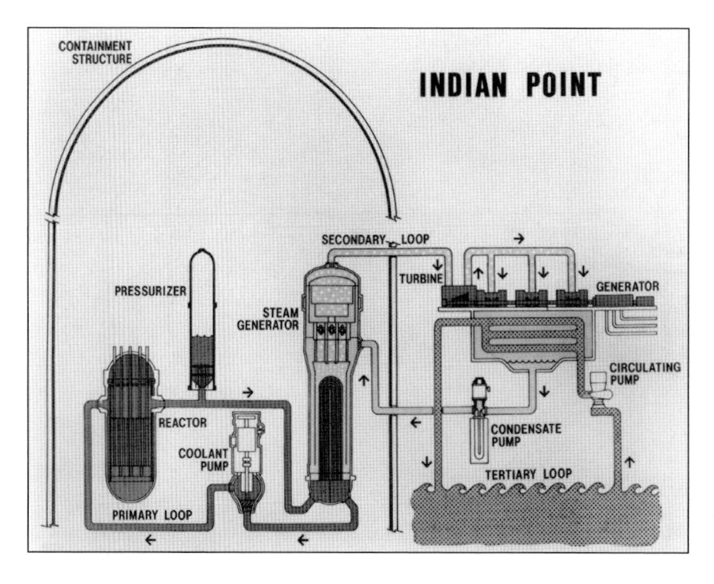

Shown is a simplified diagram of Indian Point's pressurized water reactor and main turbine generator systems. The fuel in the core heated the primary loop to about 600 degrees. That in turn heated the secondary loop into main steam inside the steam generators. The main steam then rotated the turbine generator at 1,800 RPM, causing the generator to produce 22,000 volts for Con Edison's electrical distribution system. (Courtesy of Entergy.)

Two aerial views of Indian Point are seen here. Above, the plant's situation on a zigzag on the Hudson River resulted in the plant facing northwest. Midtown Manhattan, 36 miles away, can be seen through the gap in the mountains in the background. Below, the entire site is visible. From top to bottom are IP3, IP1, and IP2. The long turbine building housed the main turbine generators for IP1 and IP2. IP3's turbine building stands alone. The L-shaped training building is at top left. The Independent Spent Fuel Storage Installation (ISFSI) is at bottom center. The Generation Support Building (GSB), built in 2003 to make room for Plant Management, Engineering, IT, and Administration, is on the left. (Both, courtesy of Entergy.)

During the operation of IP1, public tours were provided. A visitor's center was located on the hillside behind and overlooking the plant. It included tour guides, informational brochures, a movie room, and a viewing platform. People could also enter the plant and see the IP1 control room from a visitors viewing room, seen above in June 1971. The partition to the right separates the IP1 control room from the under-construction IP2 control room. Below, the same view 50 years later shows a significantly different room. The partition was removed when IP2 was completed. The visitor's viewing room has been replaced by the IP2 shift manager's office (glass enclosure). The control room supervisor was stationed in the area in front of the enclosure. The end of the IP2 supervisory panel is in the background. (Above, courtesy of Con Edison; below, photograph by the author.)

Above, reactor operators Ryan Pelky (left) and Nick Chase stand watch in the IP2 control room in 2020. The flight panel is on the right, where the reactor, turbine, and steam generator controls were primarily located. The megawatt meter typically indicated approximately 1,020 megawatts electric. The supervisory panel is behind them and controlled many of the remaining plant systems such as reactor coolant pumps, emergency core cooling, auxiliary feedwater, chemical and volume control, and the circulating and service water systems. Below, the long-retired panel of the IP1 control room can be seen behind Nick Chase. Operations typically worked a five-section rotation. The day shift, night shift, and training schedule repeated every five weeks. There were five crews, alpha through echo. IP2's crew 2A rotated with IP3's 3A, and so forth. (Both photographs by the author.)

UNITED STATES
NUCLEAR REGULATORY COMMISSION
REGION I
631 PARK AVENUE
KING OF PRUSSIA, PENNSYLVANIA 19406

Docket No. 55-5991

JUL 18 1984

Anne M. Wynne

Pursuant to the Atomic Energy Act of 1954, as amended, the Energy Reorganization Act of 1974 (Public Law 93-438), and subject to the conditions and limitations incorporated herein, the Nuclear Regulatory Commission hereby amends your license to direct the licensed activities of licensed operators at, and to manipulate all controls of the Indian Point Station, Unit No 2, Facility Docket No 50-247.

Your License No. is SOP-10161. The Docket No. is 55-5991. The effective date is May 29, 1984. The date of this amendment is June 29, 1984. This license is amended to include the correct spelling of your name.

This license is subject to the provisions of Section 55.31 of the U. S. Nuclear Regulatory Commission's regulations, Title 10, Code of Federal Regulations, Chapter 1, Part 55, with the same force and effect as if fully set forth herein.

In directing the licensed activities of licensed operators and in manipulating the controls of the above facility or facilities you shall observe the operating procedures and other conditions specified in the facility license which authorizes operation of the facility or facilities.

The issuance of this license is based upon examination of your qualifications, including the representations and information contained in your application for license filed under the docket number indicated above.

Unless sooner terminated, this license shall expire two years from the original effective date.

A copy of this license has been made available to the facility licensee.

For the Nuclear Regulatory Commission

Edward G. Greenman, Chief
Projects Branch No. 1,
Division of Project and Resident Programs

cc:
Consolidated Edison Co. of New York
Attn: Plant Manager
Indian Point Station
Broadway and Bleakly Streets
Bushanan, New York 10511

The reactor operator (RO) and senior reactor operator (SRO) license was a highly regarded credential at Indian Point and all US nuclear power plants. After over a year of training, exams were administered by the US Nuclear Regulatory Commission. An SRO was required to always be in the control room along with two ROs. Pictured above is IP2's licensed operator class in 1984. From left to right are (first row) Con Edison CEO Arthur Hauspurg, Leon Rafner, Tom O'Dell, Anne Wynne, Tom Healy, Jon Mansell, and Nuclear Regulatory Commission regional administrator Dr. Thomas Murley; (second row) George Liebler, Gary Hugo, Charlie Hock, Rich Davisson, Kevin Brooks, and George Keller; (third row) John McAvoy, Pete Schoen, Joe Poplees, Tom Brooks, Doug Eccleston, John Goodale, and Joe Goebel. John McAvoy became the Con Edison CEO in 2013. Anne Wynne was the first female senior licensed operator at Indian Point. Her license is at left. (Above, courtesy of Con Edison; left, courtesy of Anne Wynne.)

The IP2 (above) and IP3 (below) main turbine generators (MTG) are seen here. Both views are looking northwest from the generator end of the machines. These MTGs were built by the Westinghouse Electric Corporation. The three low pressure turbines are seen extending away from the main generators. The smaller units connected to the generators on the left are the exciters. In 1985, the IP2 main generator began experiencing high vibrations and required eventual replacement. The only generator of this size available was a General Electric (GE) unit. The installation of the GE machine in 1986 required significant modifications to the plant due to differences in its cooling system, and that it was designed to rotate in the reverse direction. (Both photographs by the author.)

Above, the IP3 "front standard" is seen in 2020. Here, the four 28-inch main steam lines supplied steam to the high pressure (HP) turbine (center). After leaving the steam generators in the containment building, each line traveled down and across the steam bridge into the turbine building. At full power, over 13 million pounds of steam per hour at about 750 pounds per square inch flowed to the main turbine. Each line had a stop valve and a control valve inline. The front standard is where a complex oil system controlled the position of the control valves and thus the amount of steam entering the HP turbine. This ultimately controlled the output of the main generator. Below is a cutaway of the Indian Point main turbine generator. (Above, photograph by the author; below, courtesy of the Westinghouse Electric Company LLC.)

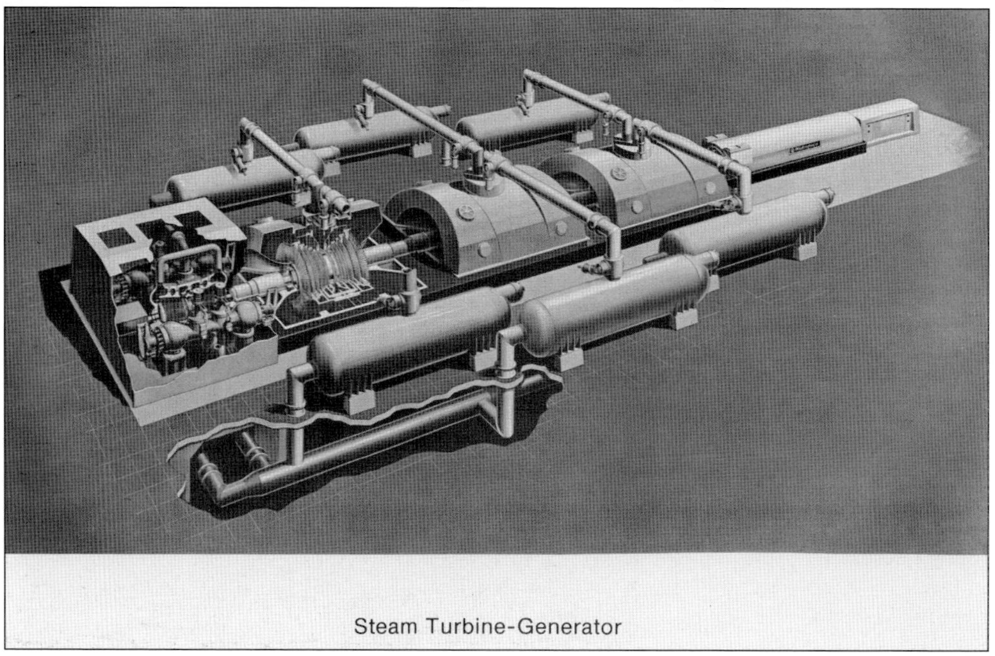

Steam Turbine-Generator

The IP3 control room was physically separate from the combined IP1 and IP2 control room. At most US nuclear plants, the major control rooms share one large room. At Indian Point, the two control rooms were about 700 feet apart. In 2011, reactor operator Ian McElroy monitors the supervisory panel on the left. In the middle, reactor operator R.T. Thomas monitors the flight panel. The megawatt meter typically indicated approximately 1,070 megawatts electric. The control room supervisor is the author, Brian Vangor, whose *Star Trek* dream had been realized. (Photograph by the author.)

Shift technical advisor Donald Vinchkoski monitors the flight panel in the IP3 control room in the mid-1980s. Shift supervisor Dick Sporbert is in the background. The shift technical advisor (STA) position was created and mandated by the US Nuclear Regulatory Commission following the accident at Three Mile Island on March 28, 1979. Each STA had a bachelor's degree in a scientific or engineering discipline and received specialized training. (Courtesy of the New York Power Authority.)

The group above was a large part of the driving force behind IP3 during the NYPA years. From left to right are (seated) Steve Munoz, Bill Josiger, Rich Ruzicka, Bob Hansler, Charlie MacKay, and Ed "Tag" Tagliamonte; (standing) Don Vinchkoski, Noel McElroy, Charlie Gorges, Charlie Caputo, Dick Sporbert, Bob Thomas, Brian Vangor, and Ed Diamond. Tag was the only person to have had an operator's license on all three IP units. He ran IP3 Operations in his final years. When anyone first went to work for him, they were scared to death of him. But eventually, they learned to love the man. Below, both IP2 and IP3 had two steam-driven main boiler feed pumps each (IP3's No. 32 is seen here). They pumped the huge volumes of feedwater back into the steam generators after the main steam had been condensed back into water. (Both photographs by the author.)

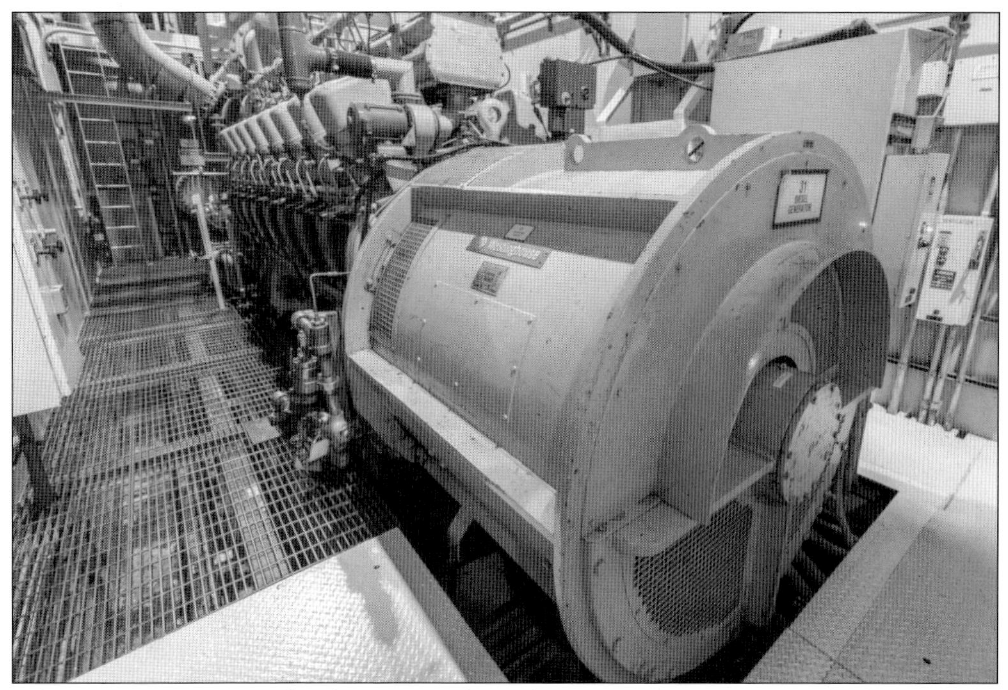

Indian Point's emergency diesel generators (EDGs) were its primary source of backup power in the event of a loss of offsite power. Each unit had three of these Alco 16-cylinder, 1,750-kilowatt, 480-volt diesel generators. The EDGs would start automatically upon a loss of offsite power and be up to rated speed and voltage in 10 seconds. They would supply power to essential equipment needed for the safe shutdown of the reactors. (Photograph by the author.)

One of IP3's main transformers is seen here in the transformer yard. Both IP2 and IP3 had two main transformers each. The 22,000-volt output of the main generator is supplied to the transformers through the ductwork seen on the left. The voltage is stepped up by these transformers to 345,000 volts to reduce losses that occur while being transmitted over long distances. (Photograph by the author.)

Both IP2 and IP3 had a steam-driven auxiliary boiler feed pump (in addition to two motor-driven pumps). These pumps could supply water (auxiliary feedwater) to the steam generators in the event of a total loss of electrical power, thus allowing the cooldown of the reactor to a safe condition. All auxiliary boiler feed pumps could be controlled from the control room or locally. (Photograph by the author.)

Circulating water pumps 34, 35, and 36 are seen here at IP3. Each unit had six of these pumps, which provided the cold river water needed to condense the main steam exhaust from the main turbine. Each could pump enormous amounts of water through its 84-inch discharge lines. The speed of these pumps could be varied depending on the time of the year and river water temperature. (Photograph by the author.)

Both IP2 and IP3 had a waste disposal panel in the Primary Auxiliary Building. IP3's is pictured here. The nuclear plant operator (NPO) could monitor various tank levels and pressures and coordinate the movements of water and gas with the control room. All NPO watch stations (typically four per unit) had local control of certain secondary equipment in their area. (Photograph by the author.)

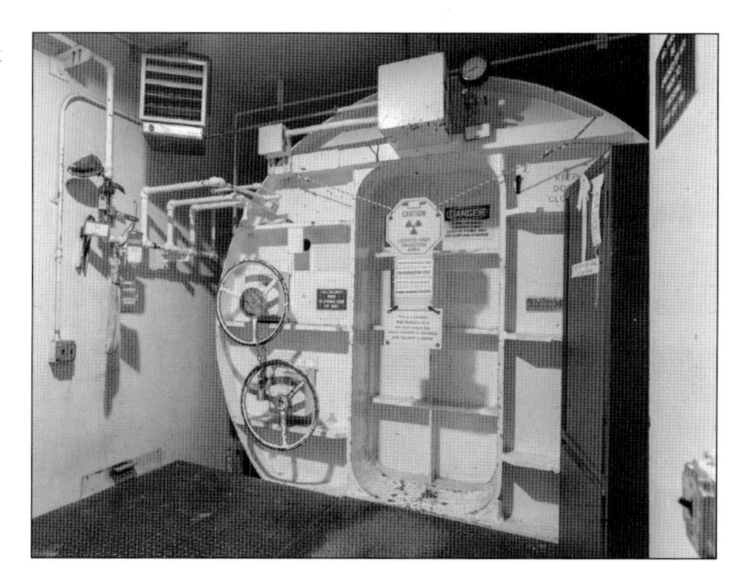

The IP3 personnel airlock into the containment building was also known as the 80-foot airlock (at IP2 as well). The outer door is seen here. The inner door was at the opposite end of the tube. The mechanism only allowed one door to be opened at a time. The handwheels on the left controlled the doors. Entries into containment when the reactor was at power were infrequent. (Photograph by the author.)

Celebrating IP2's 1978 record run of 118 days are, from left to right, Tony Nespoli, Jack O'Neill, Eugene McGrath, Tom Schmeiser, John Quirk, William "Bull" Watson, Nick Altomare, Mike Shatkouski, Max Hughes, Bill Lettmoden, Mike Mueller, Tom Lyons, Larry Townsend, Harry Morrison, Mike Anderson, and Steve Dziadik (kneeling). McGrath went on to become Con Edison CEO in 1990. As can be seen below, this record would pale in comparison to later IP record runs. IP2 operated continuously from June 1993 to February 1995 for a 616-day world record run. Westinghouse celebrated the event with candy bars for all IP2 employees. The candy bar wrapper is seen here. (Above, courtesy of Bill Durr; below, courtesy of Joe Goebel.)

Maintenance electricians Dan Knight (left) and Al Pavone perform a preventative maintenance procedure on a Westinghouse DB-50 breaker. Unlike the much more common 480-volt breakers at Indian Point, this is a 260-volt IP3 reactor trip breaker. The Maintenance Department, consisting of mechanics, electricians, and machinists, was responsible for the preventative and corrective maintenance of thousands of valves, pumps, motors, fans, and heat exchangers. (Photograph by the author.)

Instrumentation & Control (I&C) is an enormous part of any power plant. At Indian Point, thousands of transmitters, bistables, relays, power supplies, and other equipment were all wired to provide indication, control, and protection. From left to right, I&C technicians Brian Woeckener, Bob Thoma, Larry English, and John Tramm work on main generator protection circuitry in the IP2 control room. (Photograph by the author.)

The Radiation Protection (RP) Department worked around the clock protecting IP's personnel from the effects of ionizing radiation and keeping exposure as low as reasonably achievable (ALARA). Members of RP and Chemistry seen here are, from left to right, Eileen Adams, John Premich, Gregg Gross, Jose Rodriguez, Quandra Brown, John Daniele, Ray Fucheck, Frank Mitchell, Ed Hoolahan, Pete Lowe, Damien Budds, and Jeff Stewart. (Photograph by the author.)

HP1 was the entrance into the radiologically controlled areas (RCA) of IP1 and IP2. At this control point, workers logged into and out of the RCA and received instructions from the watch RP technician (behind the window) prior to entering. Here, Rich Jones Jr. is logging out of the RCA while supervisor coordinator Rich Meyer looks on. (Photograph by the author.)

The IP2 and IP3 spent fuel pools were filled with approximately 400,000 gallons of borated water each. Approximately 1,300 eight-inch-by-eight-inch-by-thirteen-foot-long fuel assemblies could be stored in each pool. IP3's pool is seen here. The boron in the water acted as a neutron absorber. The storage racks (where the tops of the fuel assemblies can be seen) are about 25 feet below the floor elevation. (Photograph by the author.)

Rad Waste technicians Tim Gemmell (left) and Grimaldy Ferreira move a drum shield in the IP2 Primary Auxiliary Building. Rad Waste performed water processing, filter changeouts, and resin movements and was responsible for the overall cleanliness of the radiologically controlled area. (Photograph by the author.)

Both IP2 and IP3 had control room simulators for initial training, retraining, exams, drills, and event simulation. Connected to specialized computers, these facilities were identical to the actual control rooms and could be used for scenarios of plant startups, shutdowns, and a wide variety of abnormal events. Instructors controlled the simulators and evaluated the operator's performance. The scenarios could be frozen, replayed, or altered to enhance student understanding. Above is the view from the instructor's booth in the IP2 simulator. Below is the IP3 simulator. The instructor's booth is behind the window on the right. (Both photographs by the author.)

In the IP3 simulator, senior reactor operators Paul Bowe (left) and Rich Ceglio discuss a system diagram with simulator instructor Charlie Gorges. Procedures, drawings, and electrical and logic diagrams were a way of life for plant operators, engineers, and technicians. Depending on a worker's discipline, everyone received training. Plant operators spent every fifth week in training. Licensed operators were required to be able to draw certain systems from memory. (Photograph by the author.)

Reactor operator and NPO instructor Lou Merlino dons the proper dress for the radiologically controlled area (RCA) and the proper personal protective equipment (PPE) while demonstrating the operation of a manual valve on one of the several mockups in the training facility. NPOs were non-licensed operators who manned certain watch stations in the plant and performed numerous tasks in the field in coordination with the control room. (Courtesy of Entergy.)

Above, fire marshal Scott Bianco instructs the Indian Point Energy Center Fire Brigade during a training exercise at the Rockland County Fire Training Center in Pomona, New York. The fire brigade was comprised of Operations personnel and was required around the clock at Indian Point. A brigade consisted of a fire brigade leader and four other fire brigade members, all of whom participated in annual retraining and unannounced fire drills. Below, fire protection supervisor Steve VanBuren instructs the brigade on the advantages of a fluoroprotein foam attack on a class B fire during annual retraining. Both VanBuren and Bianco were responsible for establishing and maintaining Indian Point as an industry leader in fire response. (Both photographs by the author.)

Dry cask storage is a method of removing spent fuel assemblies from nuclear plant spent fuel pools to provide space for additional offloads from the reactor. At right, fuel handling technicians Kevin Foley (left) and Alex Fucheck load a fuel assembly into a multipurpose canister (MPC) at IP2 in 2021. Thirty-two fuel assemblies are loaded into each MPC. Once loaded, the MPC is seal welded shut, the water is removed from the canister, and it is thoroughly tested. Following this, the MPC is placed into a concrete cask called a HI-STORM and moved to an outside storage location. Below, fuel handling technician Art Deronda drives the vertical cask transporter (VCT) with a new HI-STORM attached. The HI-STORM will be moved into the IP2 Fuel Storage Building, where the MPC will be placed inside. (Both photographs by the author, courtesy of Holtec International.)

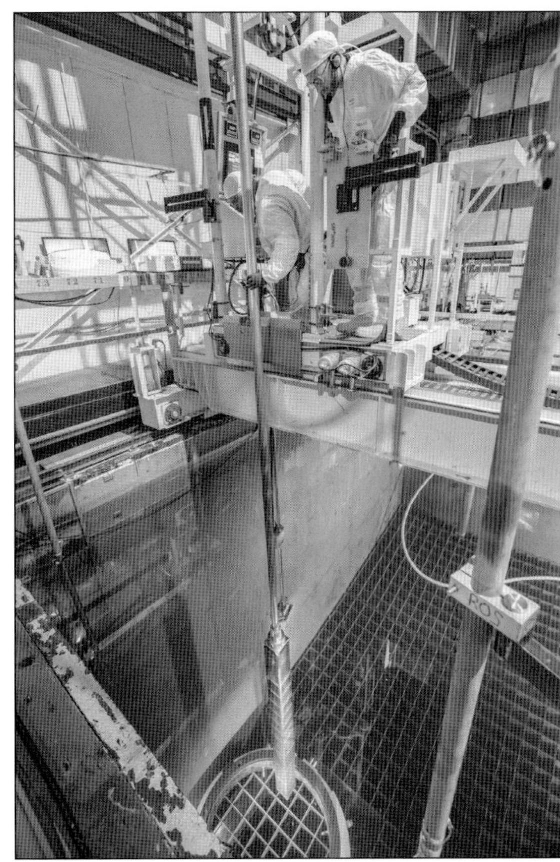

Above is the Indian Point ISFSI (independent spent fuel storage installation) pad in November 2021. Fifty-seven loaded HI-STORM casks were loaded and placed here starting in 2007. Cask 58 is empty and was being lifted by the vertical cask transporter (VCT) to begin its journey to the IP2 Fuel Storage Building for loading. Below, after being loaded, Cask 58 is pictured behind the Dry Cask Storage team. From left to right are Andy Abshire, Chris Garwood, Rich Motko, Kevin Foley, Sean Murray, Brian Woeckener, Jerzey Kopacz, Gary Couch, Bill Meyer, Gerry Fricker, Rich Jones, Jim Deronda, Steve Lucas, Chris Menzzasalma, Jeff Stewart, Alex Fucheck, Levis Alvarez, and Brian Vangor. (Both photographs by the author, courtesy of Holtec International.)

Three Indian Point site vice presidents are pictured here; from left to right are Tony Vitale, Fred Dacimo, and Rich Burroni. Vitale led the station from prior to the shutdown announcement through the actual shutdown. Dacimo oversaw the merger of the two units and then spearheaded license renewal. Burroni leads the decommissioning effort. (Courtesy of Rich Burroni.)

During its operating lifetime, up to three generations of families worked at Indian Point. In addition, many had grandfathers or uncles who worked on its construction. Lifelong Buchanan and Verplanck residents Gail and Mike Ruh each spent over 40 years at Indian Point. Son Tim followed in their footsteps. All three worked in the Operations Department. (Photograph by the author.)

Seen above in the IP3 control room in the mid-1990s, senior reactor operator and shift manager Steve Bridges is the control room supervisor. Below, in 2021, Steve's daughter and senior reactor operator Johanna Flores performs in the same capacity. Johanna and Marie Gillman were the only two female licensed operators at IP3. Johanna's husband, Gabe, was a nuclear plant operator at IP2. Anne Wynne, Catherine Cole, Deirdre (Murphy) Primrose, Michele (Savino) Hanebuth, and Melissa Streiber all were licensed operators at IP2. Gene Primrose (Deirdre's husband) and the late Harry Primrose (Gene's brother) were also licensed operators at IP2. (Both photographs by the author.)

At Indian Point, Entergy employees were recognized for outstanding achievements and contributions toward "Excellence in Nuclear Power Generation." They were chosen by their coworkers who had previously received the honor. Retired US Navy master chief and Indian Point dry cask storage superintendent Tim Salentino received such recognition. (Photograph by the author.)

Entergy purchased IP3 from the New York Power Authority in 2000. The following year, Entergy purchased IP1 and IP2 from Con Edison. Then, on April 12, 2002, after over 20 years of being completely separated, the fence came down amid much ceremony. Longtime employees walked to the opposite unit for the first time. The building of one cohesive team at the newly named Indian Point Energy Center had begun. (Courtesy of Entergy.)

All four owners of Indian Point—Con Edison, NYPA, Entergy, and Holtec International—provided philanthropic and financial support to the surrounding communities. For many years, Entergy (and Holtec International in 2021) was the major sponsor of the very popular Great Jack O'Lantern Blaze, held annually at Van Cortlandt Manor in Croton-on-Hudson, New York. (Photograph by the author.)

An employee bulletin board monitor at Indian Point shows another example of Entergy reaching out to the community. In 2014, Entergy sponsored the installation of new outdoor lighting for the historic Putnam County Courthouse in Carmel, New York. Entergy employees reached out as well. They participated in blood drives and donated to Easter Seals facilities and functions and many other charity efforts. (Photograph by the author.)

Nuclear plant operator and dry cask storage technician Mark Pasquale single-handedly spearheaded the annual holiday fundraiser at Indian Point for Rosary Hill in Hawthorne, New York. This facility is dedicated to the care of terminally ill cancer patients. Pasquale played Santa for 30 years at Indian Point and began collecting over 40 years ago. His father, Augustine Pasquale, started Christmas Gifts for Rosary Hill in 1961. (Photograph by the author.)

Entergy provided financial support to many organizations in the Hudson Valley, including the emergency room at the Hudson Valley Hospital in Peekskill. Other institutions included Phelps Memorial Hospital, the Cortlandt Volunteer Ambulance Corps, the Hendrick Hudson Free Library, Westchester Parks, Feeding Westchester, the Westchester County Library System, Historic Hudson Valley, and Arts Westchester. (Photograph by the author.)

Employee Appreciation Days were held onsite to celebrate successful outages and other milestones at Indian Point. Entergy took every opportunity to recognize its employees. Here, the IPEC Band plays at one such event in October 2016. From left to right are Frank Matra, George Seminara, Dwayne Sullivan, Jack Kenny, and Kim Passalugo. (Photograph by the author.)

Indian Point loved to eat. Here, dry cask storage technician Carl Weber enjoys the buffet at an Employee Appreciation Day. During outages and normal operation, local pizza, deli, and Italian and Chinese food establishments delivered food frequently to the main gate. During holidays, food was provided by the company day and night. Some crews cooked elaborate holiday dinners for themselves. Thanksgiving was especially memorable. (Photograph by the author.)

Seen here in the Unit 3 control room is a celebration of IP3's 40th anniversary. From left to right are Charlie Mule, Dave Daly, Mike Tesoriero, Gail Ruh, Rich Burroni, RT Thomas, Ron Carpino, Tom Ras, Brian Vangor, Nick De Vries, John Ferrick, Eugene Vanderbeek, Marc Podolski, Mike Khadabux, Tom Murphy, Tom Cramer, John Kirkpatrick, Brian McCarthy, and Mike Carroll. The cake was a "double temptation" from the Homestyle Bakery in Peekskill—a plant favorite. (Photograph by the author.)

These monoliths stood at Indian Point for almost 50 years. They described the historical features and significant events that took place in the Hudson Valley from the days of the early settlers. This exhibit, part of the Energy Education Center at Indian Point, was enjoyed by visitors, employees, and students on class trips. (Photograph by the author.)

Security at Indian Point was extremely robust. After September 11, 2001, significant enhancements were made to equipment, training, and procedures. Security was the largest department at Indian Point, and its personnel were highly trained professionals. Security officers Ted Gordon (left) and Steven Stropoli perform a vehicle search at the plant's main entrance. (Photograph by the author.)

Indian Point's emergency plan provided detailed instructions for the actions to be taken by the plant staff if one of the four emergency classifications was ever declared. The emergency director (ED) and staff would be stationed here at the emergency operations facility (EOF), where they would manage the event while in constant communication with the control room and all county, state, and federal agencies. (Photograph by the author.)

Five

OUTAGES

Originally, like most nuclear plants in the United States, IP2 and IP3 operated with fuel cores that would last for 18 months. At the end of a cycle, the plant would be shut down, disassembled, and refueled for the next cycle. Hundreds of other maintenance activities took place as well. In the early days, refueling outages could last up to 100 days. Over time, two things changed—the fuel was enhanced so that a cycle could last for two years, and the duration of a refueling outage was eventually reduced to about 30 days as scheduling, planning, and resources improved.

Typically, about half of the 193 fuel assemblies in the reactor core were replaced with new fuel assemblies during each refueling outage. Each fuel assembly spent approximately four years in the reactor. Indian Point established operating cycles where IP2 refueled every spring on the even years and IP3 every spring on the odd years. Each outage had an identifier (2R15 would mean Unit 2's 15th refueling). During their lives, IP2 operated for 24 cycles, while IP3 operated for 21 cycles.

Outages, whether planned or unplanned, were all high-activity, around-the-clock, all-hands-on-deck events. Each activity was scheduled down to the minute. For planned outages, hundreds of nuclear worker contractors from around the country were brought in to supplement the permanent workforce. They found local temporary housing or lived in their trailers. The Access and Training departments worked tirelessly to obtain the necessary personnel paperwork and provide these individuals with the required training and exams necessary for entry. Crews of technicians from Westinghouse performed reactor refueling, reactor vessel head testing, and steam generator, reactor coolant pump, and turbine generator maintenance. Additional radiation protection technicians, laborers, carpenters, electricians, fitters, millwrights, and boilermakers were also required. The parking lots overflowed. Temporary trailers were set up as 24-hour offices and break rooms. Take-out food runs were constant and local businesses thrived. The work got done, and the unit was returned to service.

The sign that a refueling outage was imminent was the arrival of these canisters from Westinghouse in Columbia, South Carolina. Each canister held two new fuel assemblies. With eight to twelve canisters arriving every week, it took approximately one month just prior to the outage to receive and store the 60 to 96 new fuel assemblies required for refueling. (Photograph by the author.)

Senior radiation protection technician Sonya Slivinski uses a neutron meter to check for any indication of radiation emanating from the new fuel assemblies in the just-opened canister. Alpha and gamma dose rates were measured, as were smears taken to check for contamination. New fuel assemblies were handled with great care and were only touched using white cotton gloves. (Photograph by the author.)

In the IP3 Fuel Storage Building truck bay, from left to right, refueling technicians Bob Parks, Kevin Shoptaw, and Pete Pliska work on loosening the clamping frames that secure the fuel. The assemblies will be rotated vertically on pivots on the right. New fuel assemblies were supported in the canister by a shock-mounted frame and had seismic accelerometers to warn of an impact during transport. (Photograph by the author.)

Dry cask storage technician Jim Fandel prepares to attach the new fuel handling tool to a new fuel assembly. Once attached, the remaining clamps would be loosened, and the assembly lifted from the canister. Several people below Fandel carefully guided the assembly until it was clear of the support frame. (Photograph by the author.)

Reactor engineer Floyd Gumble inspects a new fuel assembly at IP3 in the mid-1980s. Prior to being accepted and moved to the spent fuel pool, each 1,500-pound, 13-foot-long fuel assembly was thoroughly inspected for damage or debris. The reactor engineers were also responsible for maintaining precise documentation and a record of all moves and location changes for each assembly. (Photograph by the author.)

The IP3 fuel vault is seen here filled with new fuel assemblies ready for refueling. Seventy-two fuel assemblies could be stored here. Each assembly cost approximately $1 million. Just prior to or during the refueling, each assembly was transferred to a location in the spent fuel pool. During refueling, each was moved into the containment building via the underwater transfer system and placed into the reactor. (Photograph by the author.)

In the late 1980s, the author is directing the removal of a new fuel assembly from the fuel vault so that it can be placed into the spent fuel pool in preparation for refueling. The new fuel handling tool is connected to the assembly's top nozzle. Each of the 204 fuel rods contained hundreds of 0.375-inch uranium dioxide pellets stacked atop each other. (Courtesy of the New York Power Authority.)

Westinghouse refueling technician Carl Pierce directs a new fuel assembly into the new fuel elevator. Once fully inserted, it will be released from the handling tool and the elevator will lower the assembly into the pool to the height of the storage racks below. The assembly will then be relocated to its designated storage location for refueling. (Photograph by Jon Summers.)

Dry cask storage technicians Rich Jones (left) and Kevin Slesinski prepare the IP2 fuel handling machine to move a new fuel assembly out of the new fuel elevator. Jones operated the machine and handled the fuel. Slesinski was the spotter and was responsible for verifying correct operation of the machine and correct placement of the fuel assembly. (Photograph by the author.)

This fuel assembly, 3H84, seen in the IP3 new fuel vault, was the last fuel assembly delivered to Indian Point on February 4, 2019, for IP3's final refueling. The insert, 20W0456, is a set of neutron absorber rodlets. A total of 3,998 fuel assemblies like this were used at IP2 and IP3. IP1 used 404 fuel assemblies of its own type. (Photograph by the author.)

Refueling outages typically started at midnight when the reactor was tripped (shutdown). A few hours prior to this, a containment entry brief was held to discuss entry requirements, radiological conditions, and crew responsibilities. Here, RP supervisor Bob Solanto gives the brief. From left to right, Jeff Stewart, Ed Goetchius, Jonathon Szabo, and Josh McCarty listen in. (Photograph by the author.)

This was a common scene at the beginning of every refueling outage. Workers waited in the 80-foot airlock for the operator to open the inner door (handwheel on the left) into the containment building. Officially named the Vapor Containment, or just "VC," building integrity (one door open at a time) was required to be maintained until the reactor was cooled to below 200 degrees Fahrenheit. (Photograph by Jon Summers.)

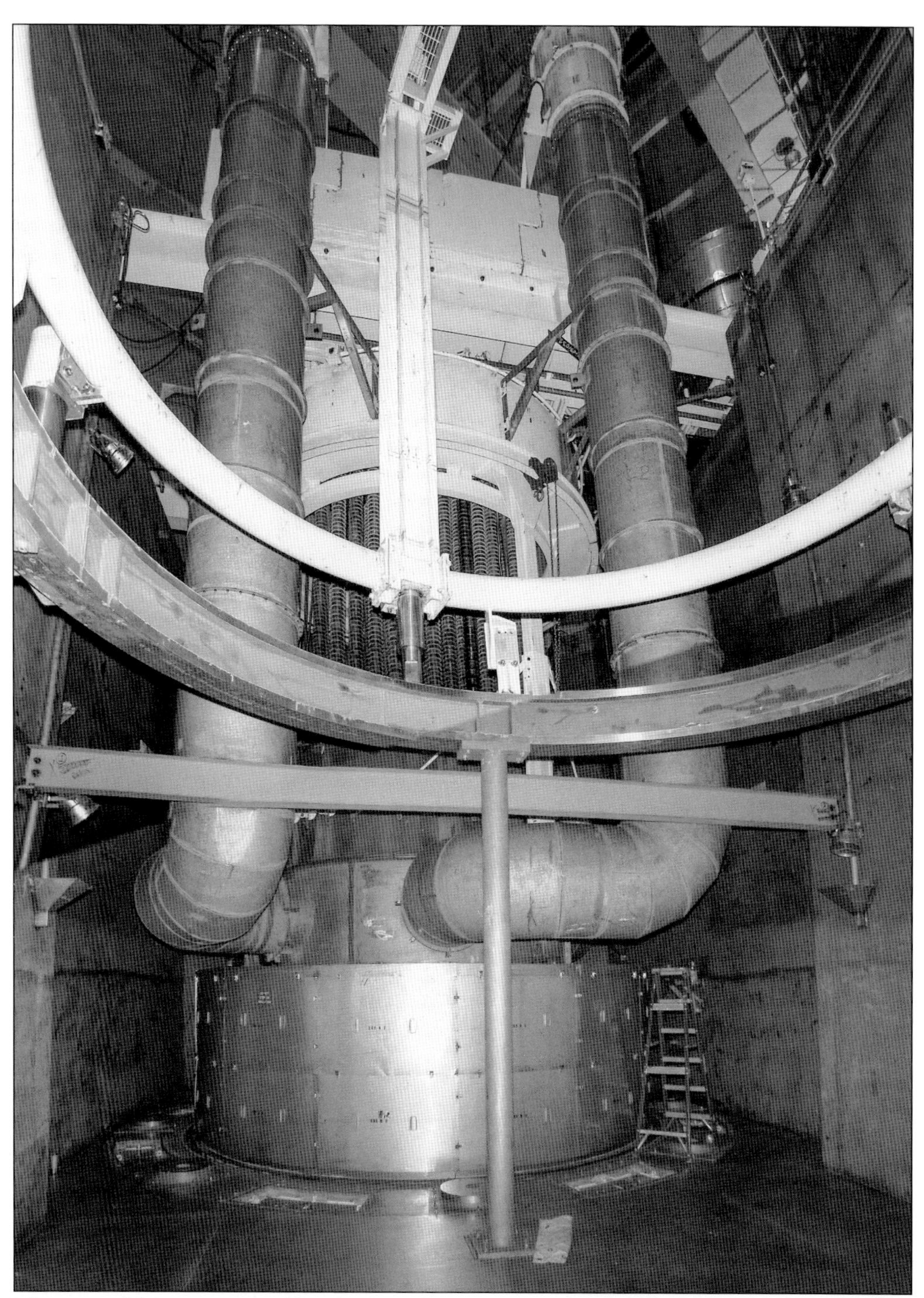

This view of the reactor vessel head is from the reactor cavity floor (69-foot elevation). The vessel head is at floor height behind the mirror insulation. The four 25-ton concrete missile blocks are on top. The control rod drive mechanisms (CRDM) and rod position indication (RPI) coil stacks are in the center. Two of the four CRDM ventilation duct risers are seen going up to the fans above. (Photograph by the author.)

A refueling outage always began with workers removing the four CRDM fans to start reactor disassembly. Here, 33 CRDM fan is about to be lifted away. Until the plant was cooled down and the equipment hatch removed, work in containment was typically hot, noisy, and uncomfortable. Crews worked in three-hour "jumps." (Photograph by Jon Summers.)

Control room supervisor and refueling supervisor Jon Summers oversees electrical disconnect work on the "bedspring" at IP3. The bedspring supported the cables that connected to the control rod drives and rod position indication systems down below on the reactor vessel head. Each refueling, the wires were disconnected, and the bedspring rotated out of the way to provide access to the vessel head below. (Photograph by the author.)

The fans, ducts, and missile blocks have been removed from atop the reactor vessel head. On the reactor cavity floor below, the mirror insulation covering the actual closure head is still installed. The reactor itself is entirely below the floor. Once the insulation is removed, the reactor vessel head studs could be accessed and detensioned (loosened) and the head lifted out of the cavity. (Photograph by Jon Summers.)

This view is looking straight down into the IP2 containment transfer canal. The cart that carried the horizontally positioned fuel assembly between the containment and fuel storage building ran on the rails and wheels at the bottom. The upender basket is at the left end of the rails, with the pivot between the two workers. The manipulator mast is on the right. (Photograph by Jon Summers.)

In this scene from the mid-1980s, the IP3 reactor vessel head is being lifted off the reactor vessel and out of the reactor cavity. Workers monitored the lift from the cavity floor. The top of the upper internals package and the control rod drive shafts will be exposed once the head is removed. Immediately after head removal, the reactor cavity was filled with approximately 24 feet of water. (Photograph by the author.)

The 60-ton upper internals package is seen here being removed from the reactor vessel and placed in its underwater storage stand. The cavity water level was increased to 93 feet, 6 inches for this move to maximize radiation shielding. Out of view below, the fuel in the reactor core is uncovered and ready for offloading. (Photograph by Jon Summers.)

This unique IP3 view from the mid-1980s is looking straight down toward the reactor cavity from the polar crane above. The reactor is in the center with the core clearly visible. The manipulator crane is at the bottom of the photograph. The upper internals package is at the top, in its underwater stand. The transfer canal to the fuel storage building is at bottom left under the manipulator crane. (Photograph by the author.)

On May 11, 2020, during IP2's final core offload, Westinghouse refueling technician Bryce Musa looks up from the manipulator crane. The shaft extending down below the crane was the mast. At the bottom of the mast was a gripper that latched onto each 1,500-pound fuel assembly. The manipulator crane rode on rails on the 95-foot elevation over the reactor cavity. (Photograph by the author.)

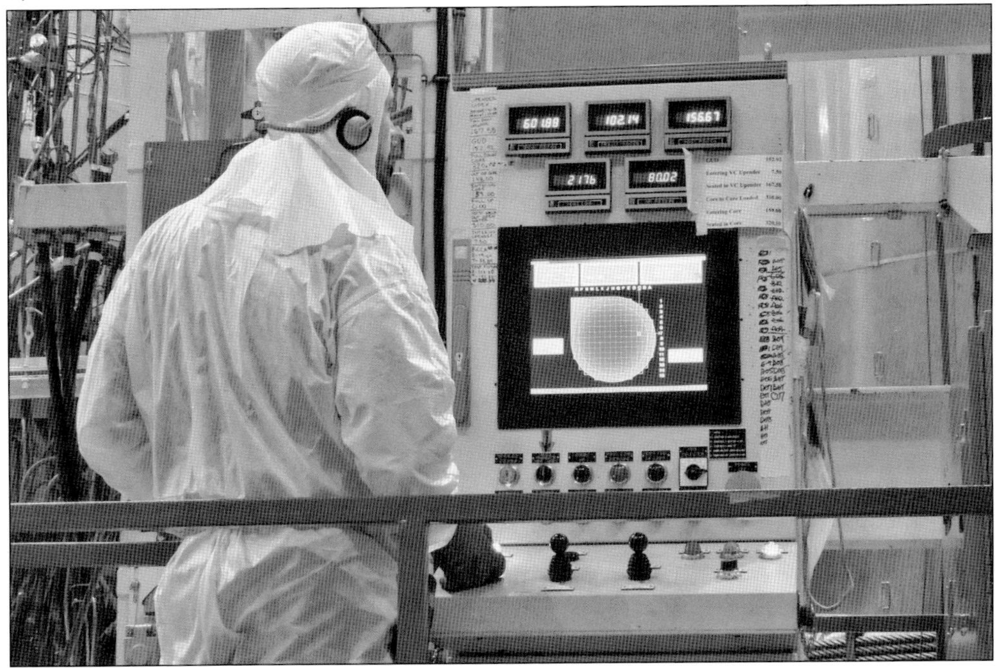

Westinghouse refueling technician Pete Pliska operates the IP3 manipulator crane during IP3's final core offload. The metal framework behind him was the support structure for the mast. The control console showed the exact location of the mast over the core and contained every indication and control the operator needed to move a fuel assembly into or out of the core. Typically, a core offload or core reload took about 50 hours to complete. (Photograph by the author.)

During refueling, all movement of fuel had to be coordinated with the control room, which had indications of conditions inside the reactor. Constant communication between the control room, manipulator crane, and fuel storage building was required. The control room gave permission for each core alteration (fuel movement). Here, IP3 reactor operator Ian McElroy directs the movements. (Photograph by Jon Summers.)

Refueling technicians over the IP3 spent fuel pool move a fuel assembly just removed from the core to its storage location. The fuel storage building (FSB) was connected to containment via an underwater transfer tube. Each side had a fuel upender basket that rotated the fuel from vertical to horizontal. Once horizontal, the basket moved through the transfer tube on a wheeled cart driven by a motor and underwater cables and sheaves. (Photograph by the author.)

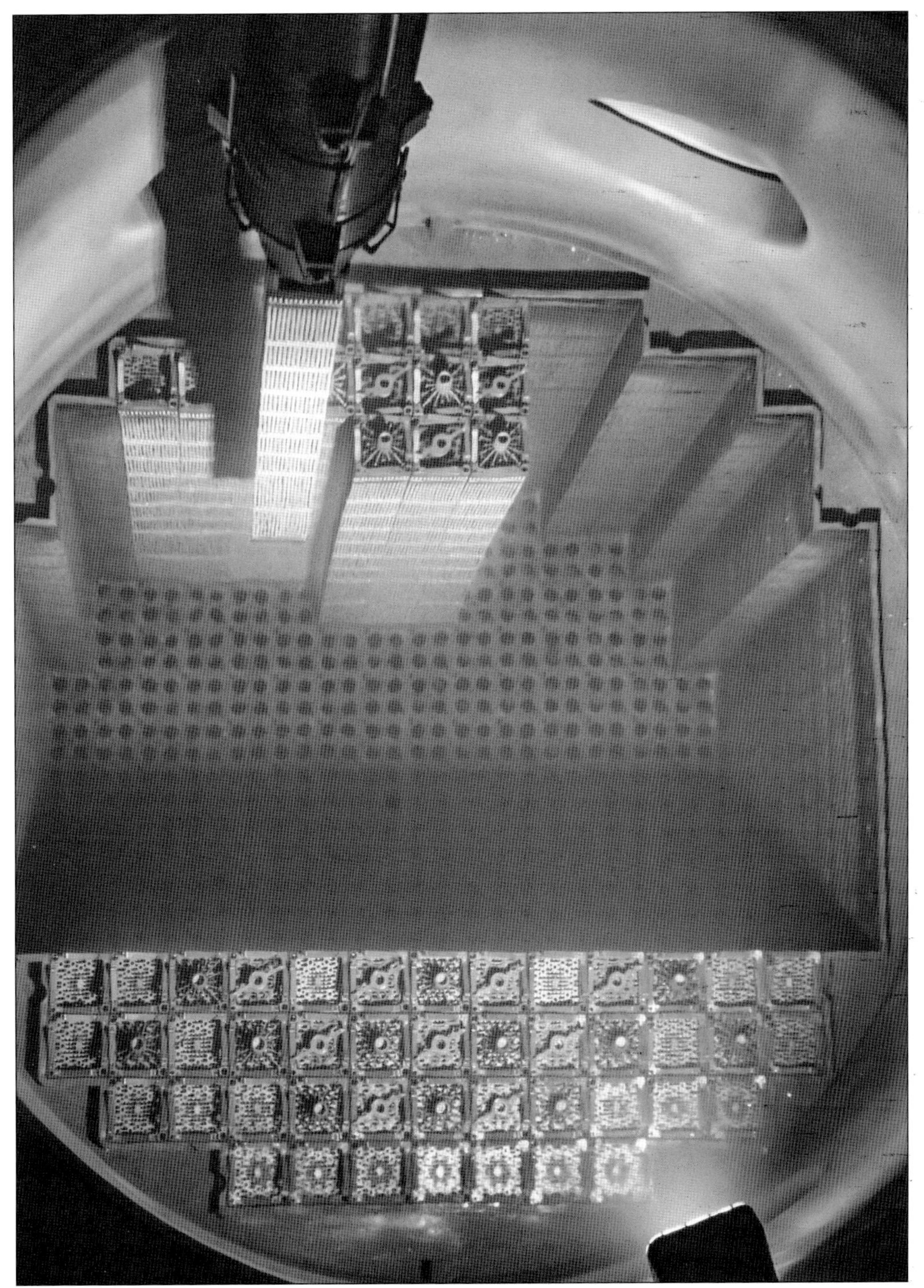

This 2021 view looks straight down into the IP3 reactor core during 3D21 (Unit 3 defueling cycle 21). This was the final defueling of the IP3 reactor. A fuel assembly is being lifted by the manipulator crane. The lower core plate, where the reactor coolant entered the core and where the fuel assemblies sat, is at center. (Photograph by the author.)

Pictured here from left to right are refueling senior reactor operator Russ Long, Master-Lee manipulator crane operator Tim George, and containment coordinator Vinny Coulehan on the IP3 manipulator crane. They are positioned over the transfer canal awaiting the arrival of the next fuel assembly from the fuel storage building during IP3's final core reload on March 29, 2019. (Photograph by Jon Summers.)

Shift manager Nick Lizzo (standing) and control room supervisor Matt Johnson discuss plant status in the IP3 control room during a refueling outage. Hundreds of activities and tests had to be monitored by the control room to ensure the safety of the workers and the proper configuration of the plant. External to the control room, the Outage Control Center monitored the status of the progression of work and provided resources to resolve issues. (Photograph by the author.)

RCP maintenance was performed during most outages. Personnel from the plant and from Westinghouse teamed up to do this work. Here, an RCP motor on the 95-foot elevation of the IP3 containment awaits installation. These 37-ton motors were rated at 6,000 horsepower. Each of the four RCPs pumped almost 90,000 gallons of water through the reactor every minute. (Photograph by Jon Summers.)

In the late 1970s, workers take a break while bringing an RCP motor into the IP2 containment building. Work in containment (a radiologically controlled area) was at times exhausting. To prevent the ingestion of contamination, there was no eating, drinking, or chewing. There were no bathrooms, and workers were not allowed to scratch their faces. (Courtesy of Con Edison.)

For many, outages meant life in a trailer. Contractors were given onsite trailers to occupy. Plant supervision moved to the trailers to be in the same location. The trailers contained computers, parts, procedures, briefing areas, refrigerators, microwaves, food, coffee, and snacks, especially chocolate. From left to right, refueling supervisors Tom Pasko and Tim Salentino and Westinghouse's Steve Harms monitor the refueling. (Photograph by the author.)

The turbine hall was a beehive of activity during most outages. Maintenance was performed on the turbines, moisture separator reheaters, and the main generator itself. Several companies and multiple craft unions performed the work. Specialized equipment for this type of maintenance was brought in including sand blast tents, power rollers, and a job-specific tool room. (Photograph by the author.)

A low pressure turbine spindle and its last row of blades are seen here following the removal of the outer casing. Steam entered each LP in the center and flowed in both directions toward each end and ultimately into the exhaust openings into the condenser. A turbine bearing and the turbine shaft are seen in front of the worker. (Photograph by Jon Summers.)

The massive size of an LP turbine spindle is evident in this photograph. As the steam traveled outward from the center of the spindle, its pressure dropped. This required the blades to be sequentially larger going outward to provide an equal rotational force on the turbine shaft. During outages, blades were inspected, repaired, or replaced. (Photograph by Jon Summers.)

After experiencing high vibrations in 1985, the original Westinghouse IP2 main generator was replaced in 1986 by a unit from General Electric. At the time, this unit was the only one available at the size required. Here, on October 22, 1986, the new generator is being lifted from the south loading well at IP2 up onto the operating deck (53-foot elevation) of the turbine building. (Courtesy of Con Edison.)

During refueling, work took place on equipment all over the plant. In addition, all outages were not refueling outages. Maintenance outages were required as well. Some equipment could be taken out of service while the plant was operating. Here, maintenance personnel remove a coupling from a circulating water pump at IP3. (Courtesy of John McCarty.)

Both IP2 and IP3 had four large transformers each—two main transformers, a station auxiliary transformer, and a unit auxiliary transformer. Transformer replacements were huge undertakings requiring heavy moving equipment. Above, one of IP3's 200-ton main transformers is being slid out of the transformer yard and under the steam bridge. Below, a replacement transformer arrives by barge at the north dock. Weeks 533 lifts a main transformer to the awaiting Goldhofer transporter. Weeks 533 could lift 500 tons. The Goldhofer had 12 rows of eight wheels and was driven by remote control. (Above, photograph by the author; below, courtesy of Con Edison.)

Initial industry practices in chemical control of the secondary feedwater caused steam generator (SG) tube issues in many US nuclear plants, including IP2 and IP3. As a result, both IP2 and IP3 replaced their original SGs with new and improved models. IP3s were replaced in 1989. This was the first time in the nuclear industry that SGs were replaced in one piece. IP2's SGs were replaced in 2000. Above, one of IP2's new SGs is being moved to a storage facility onsite after delivery. Although IP2's new SGs were built first, IP3's were installed first. Below, in 2021, senior radiation protection technician John Dorsey inspects the original IP2 SGs placed here in a "mausoleum" in 2000. (Above, courtesy of Con Edison; below, photograph by the author.)

After the core was reloaded and the upper internals reinstalled, the reactor cavity was drained of all water. Installation of the reactor vessel head was the next step. On April 9, 2018, Westinghouse refueling technician Terry Ashley directs the polar crane operator in setting the 169-ton IP2 reactor vessel head for IP2's final refueling (2R23). (Photograph by Cliff Gates.)

During the lowering process, workers climbed down into the reactor cavity to ensure that each of the 53 control rod drive shafts entered the funnels under the head. The funnels guided the shafts up into the control rod drive mechanisms. In this view from a head-mounted camera, the author, using a shepherd's hook, adjusts the position of the drive shafts. (Photograph by the author.)

Following head installation, the 54 studs and nuts were "turned in." Three hydraulically operated stud tensioning machines were used to pull the seven-inch diameter studs approximately fifty-thousandths of an inch. The nut was then tightened, and the stud was released. Here, two refueling technicians screw a tensioner onto a stud. Following tensioning and the remaining reactor reassembly, plant startup commenced. The outage was over. (Photograph by the author.)

Six

Shutdown

In 2017, Entergy voluntarily entered a shutdown agreement with New York State and environmental groups, ending years of legal proceedings over the extension of the operating licenses for both plants and the use of the Hudson River for cooling. Both units had entered a period of extended operation after the original license dates of September 28, 2013, and December 12, 2015, had passed. The agreement would close IP2 on April 30, 2020, and IP3 on April 30, 2021. Economics played a huge role as well. The increasing availability of natural gas reduced the price of wholesale electricity in the state, thus increasing competition and reducing revenues from Indian Point.

Entergy provided many options to its Indian Point employees. Initially, all were enticed to remain at the plant until the shutdown of IP3 was performed in April 2021. Additionally, employees were given future career opportunities at Entergy facilities in Mississippi, Arkansas, Louisiana, and Texas. About 350 employees were solicited to remain at Indian Point following the shutdown as several plant systems were required to remain in operation for extended periods. Many retired after 30- or 40-year careers.

As planned, IP2 performed an orderly shutdown on April 30, 2020. The reactor was defueled as part of 2D24. IP3 went exactly one year later, following a world-record run of 753 days of continuous operation. This was the first time in Indian Point's history that either of the units had operated from "breaker to breaker" (starting up after a refueling outage and operating continuously until shutdown for the following refueling). IP3's defueling, 3D21, was completed on May 12, 2021.

On May 20, 2021, Separation Day, several hundred Indian Point employees said goodbye to the plant and many of their coworkers and friends for the last time. On May 28, the sale of Indian Point to Holtec International was finalized. Holtec will now proceed with the decommissioning of Indian Point, ending 59 years of electrical generation by nuclear power in the Hudson Valley.

In accordance with the shutdown agreement entered by Entergy, IP2 was taken out of service on April 30, 2020. From left to right, control room supervisor Kurt James, shift manager John Baker, and reactor operators Dave Owen and Tom Loughran discuss shutdown procedures. A plant "coastdown" due to fuel depletion had been in progress since March 17, thus reducing power to approximately 70 percent for the shutdown. (Photograph by the author.)

This single button tripped the reactor by removing power from the control rod drive mechanisms, allowing all control rods to fall into the core. This immediately shut down the nuclear chain reaction. It also closed the main turbine's stop and control valves, shutting off the steam supply to the machine. After a 30-second delay, the main generator's output breakers opened, disconnecting the generator from the electrical grid. (Photograph by the author.)

At 11:00 p.m. on April 30, 2020, shift manager Gene Primrose pushed the IP2 reactor trip button for the last time. Primrose was the most senior licensed operator at IP2. His wife, Deirdre, was also a licensed operator at IP2. Reactor operator Dave Owen was monitoring steam generator levels. (Photograph by the author.)

IP2 control room supervisor and refueling senior reactor operator Cliff Gates was on the manipulator crane for the final fuel assembly removed from the core, 2C02 from core location H01. It was unlatched in the upender at 1:42 a.m. on May 12, 2020, completing the final defueling of the IP2 reactor. (Photograph by the author.)

Control room supervisor Kevin Brooks Jr. briefs the control room operators just before the shutdown of IP3. Shift manager John Ryan looks on. Brooks's father, Kevin Sr., was a licensed operator on IP2. Kevin Jr. told his crew, "753 days, very proud of what we accomplished." He then directed reactor operator Russ Warren, "Russ, for the last time at Indian Point Unit 3, trip the reactor." (Photograph by the author.)

From left to right are reactor operators Russ Warren, Ron Gores, and Mark Lewis. Warren pushed the reactor trip button and shut down the IP3 reactor on April 30, 2021, at 11:00 p.m. Here, seconds later, he is verifying all rods have fallen into the core. As with IP2, a coastdown to approximately 70 percent power had been completed previously. (Photograph by the author.)

The final fuel assembly, 3G58, was removed from the IP3 reactor, core location H15, on May 11, 2021. It was unlatched in the upender a few minutes later at 11:15 p.m. All three reactors at Indian Point were now defueled. IP2's spent fuel pool now contained 990 spent fuel assemblies, while IP3's contained 1,280. (Photograph by the author.)

Following the last fuel assembly being removed from the IP3 core, refueling team members signed the top of the manipulator crane mast to celebrate IP3s just-completed world-record run for a light water reactor. The date was May 11, 2021. This proud but bittersweet achievement culminated many long careers and caused many to pursue new ones elsewhere. (Photograph by the author.)

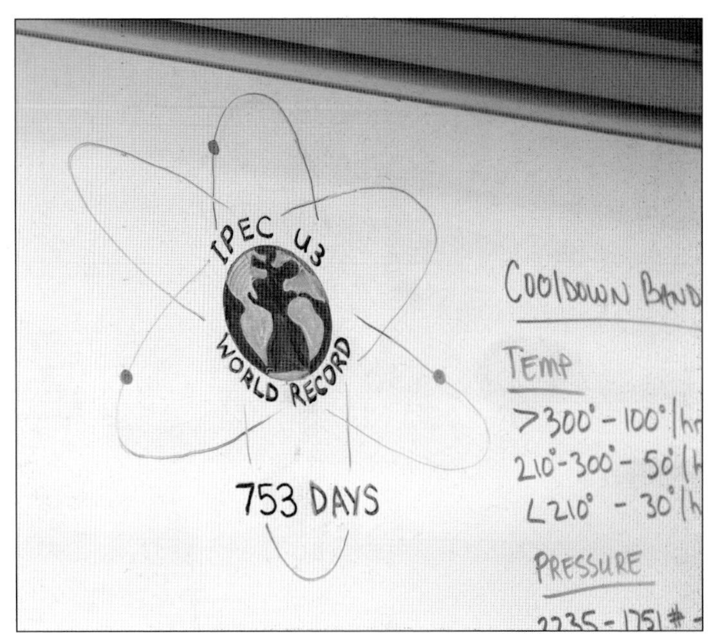

IP3 control room supervisor Mike Weeden drew this image on the control room whiteboard commemorating IP3's world-record run. For the first time, an Indian Point unit ran breaker to breaker. This 753-day continuous run occurred from April 9, 2019, to April 30, 2021. (Photograph by the author.)

For many years (and still in 2022), "Welcome to Indian Point where your safe and error-free performance will add to our legacy of excellence," was the audio message heard by every person going through the security monitors when entering the plant. It reminded everyone of where they were and what was expected of them. (Photograph by the author.)

In the weeks leading up to the final shutdown, signs sprang up from the local communities showing their gratitude for the plant's safe operation and contribution toward clean energy. Indian Point employees appreciated the recognition. Above, on April 30, 2021, at Lents Cove, the Verplanck Volunteer Fire Department posted this sign and hoisted Old Glory in recognition. Also on April 30, a celebration was held at Lents Cove to honor Indian Point workers. Below, a hard hat was hung for each employee. Indian Point always supported and worked closely with its local communities. The plant was part of the landscape for over 60 years and will continue to be for the foreseeable future. (Both photographs by the author.)

Thank you, Indian Point

For nearly 60 years, Indian Point has provided electricity that powered the lives of millions of New York City and lower Hudson Valley residents – safely, reliably, and securely.

But Indian Point's legacy is much greater than that.

The men and women of Indian Point and their dedication to professionalism, safety, and service will be remembered and celebrated as the power plant's legacy.

The people of Indian Point designed and built the plant, tested its components, and started it up for the first time.

They harnessed the energy generated by the plant around the clock, with dedicated personnel onsite, 24 hours a day, seven days a week, year after year, for the last 59 years without interruption.

They powered our lives through hot summer days and freezing cold winter nights.

The people of Indian Point raised their children, volunteered with local organizations, and lived their lives in this area, serving the community professionally and personally.

Indian Point shut down on April 30 for the last time after running continuously for 753 days since it was last refueled – a new world record for light water reactors and a fitting tribute to its people.

When I look back on Indian Point Energy Center and its history of operation, I will remember the dedicated men and women who I met, along with those who came before them, with deep appreciation and respect for having served with honor and distinction.

The people who passed through the plant will live on in the memory of everyone who has ever worked at Indian Point.

On behalf of the team,

Chris Bakken
Chief Nuclear Officer, Entergy Nuclear

Following the plant's shutdown, Entergy's chief nuclear officer Chris Bakken sent this message to Indian Point employees. Bakken clearly summarizes the feeling the author has tried to convey within the pages of this book and the feeling expressed to him by everyday people on the subject. (Courtesy of Entergy.)

The Indian Point Heritage Project was founded and directed by Dr. Michael Conrad of Clarkson University and sponsored by Entergy. Research assistant Ludmilla Edinger and web designer Joseph DePinho assisted in the project. Extensive audio interviews with Indian Point employees were recorded and posted to the project's website. The interviews, as well as photographs and videos, can be found at ipheritage.org. (Courtesy of Joseph DePinho.)

Created in 2017, *Indian Summer* is a 104-inch-by-62-inch oil on wood painting by Peekskill artist Andrew Barthelmes. From steamships to steam generators, the entire history of Indian Point is intertwined in this one mural. It not only summarizes the entire span of information presented by this book, but it is also a fitting final image. (Original painting by Andrew Barthelmes.)